WE NEED TO TALK ABOUT KELVIN

We Need to Talk about Kelvin

What Everyday Things Tell Us
about the Universe

MARCUS CHOWN

ff

faber and faber

First published in 2009
by Faber and Faber Ltd
Bloomsbury House, 74–77 Great Russell Street
London WC1B 3DA

Typeset by Ian Bahrami
Printed in England by CPI Mackays, Chatham ME5 8TD

A CIP record for this book
is available from the British Library

ISBN 978–0–571–24401–0

2 4 6 8 10 9 7 5 3 1

To Karen and Jo,
With love, Marcus

Contents

Foreword

'To see a World in a Grain of Sand
And a Heaven in a Wild Flower,
Hold Infinity in the palm of your hand
And Eternity in an hour.'
                    William Blake
            ('Auguries of Innocence')

The idea of this book is simple: to take familiar features of
the everyday world and show how, in the light of our current
scientific knowledge, they tell us profound truths about the
ultimate nature of reality; to read the cosmic signs in the
everyday world. Or, in the words of William Blake, 'To see a
World in a Grain of Sand' – or a falling leaf or a rose or a
starry night sky . . . For instance:

• The reflection of your face in a window tells you about
the most shocking discovery in the history of science: that
at its deepest level the world is orchestrated by random
chance; that ultimately things happen for *no reason at all*.

• The fact that iron is common – in the steel of the cars we
drive, the framework of the buildings we work in, even in
the blood at this moment coursing through your veins –
tells you that somewhere out in the depths of space there
must exist a blisteringly hot furnace at a temperature of
about 4.5 billion degrees.

• The fact that there are no aliens on Earth – either

loitering on street corners, flying angelically through the sky above, or materialising and dematerialising like crew members of the starship *Enterprise* – tells you . . . well, we don't actually know what it tells you. It could be that we are the first intelligence to arise in our Galaxy, possibly the whole Universe, sentenced to cosmic solitary confinement on Earth with no one else to talk to. Or it could be that the Universe is so dangerous a place that every space-faring race is wiped out before it can come our way. This is the one everyday observation where – frankly – your explanation is as good as mine.

The idea to write about what the everyday world can tell us about the Universe came to me in the publicity phase between books. Being an author is an all-or-nothing existence. Much of the time, I am locked away with only George and Reg the goldfish for company (sadly, Laura passed away during the writing of this book). For a brief time, however, when doing publicity, I get out and about and actually meet people in a whirl of sociability. And the skill required to publicise is entirely different to that required to write a book. In radio interviews, I have at most a few minutes to convey something that will lodge in the mind of listeners. In public talks, most of the audience may not have a science background. So I am continually grasping for new, visual, snappy ways of saying things. And one thing I suddenly realised while doing this – an obvious thing, really – is that, in talking to non-scientists, I tend to latch onto an everyday observation, then relate it to the deep physics it exemplifies.

At the 2008 Edinburgh Science Festival, for instance, I needed to highlight the basic paradox that leads to quantum theory, our best description of the microscopic world of atoms and their constituents. So I drew people's attention to a light bulb in the auditorium and pointed out how the light waves that emerge from it are about 5,000 times bigger than the atoms themselves. I then took a matchbox from my pocket and said, 'Say I opened this matchbox and out drove a 40-tonne truck. *That's* what it's like for light streaming out of that light bulb.'

And one day, a light bulb *did* go on in my head. I suddenly thought, 'Why don't I write a book in which each chapter takes an everyday observation of the world and points out the profound thing it tells us about ultimate reality?' Simple as that. Why had I not thought of that before? Suddenly, I could see all sorts of things I wanted to write about coming together. It was a powerful unifying thread.

I was excited. But I was also worried that I might repeat myself. I hope, however, that although I do return to things I have talked about in previous books such as *The Magic Furnace* and *Quantum Theory Cannot Hurt You*, I have deepened the discussion, shown things in a new light. A good example is the 400-year-old mystery of why the sky is dark at night. Like 99 per cent of astronomers, I used to think the blackness at midnight is telling us that the Universe has not existed for ever but was *born* – that the evidence for the Big Bang has been staring us in the face since the dawn of human history, had we only the wit to recognise it. I may even have said this in my book *Afterglow of Creation*. Now I realise that the darkness at night is not telling us that at all. Most astronomers are wrong. And, bizarrely, it was Edgar Allan Poe, of all people, who was the first person to catch a glimpse of the truth.

Another example of something I return to but elaborate on is the boundless variety of the world we live in. Ultimately, this is due to the Pauli exclusion principle, which prevents electrons piling on top of each other and, by doing so, is responsible for there being many types of atom rather than a single kind. I was aware that, in *Quantum Theory Cannot Hurt You*, I had fallen short of a complete explanation. I managed to show how nature permits two indistinguishable particles to behave in two distinct ways: to be either gregarious or antisocial. I then said that nature avails itself of both possibilities. Particles with a particular type of 'spin' turn out to be antisocial – like electrons – whereas particles with a different type of spin – such as photons – are gregarious. But what I didn't explain is, what the hell has spin got to do with what option a particle takes up? I had given only *half* the explanation. In my defence, it took Wolfgang Pauli from 1926, when he proposed the exclusion principle, until 1941 to come up with an explanation for what spin had to do with it – the so-called spin-statistics theorem. So I do not feel *that* bad. In this book, however, I hope that I have given a complete explanation, one that – as far as I know – does not exist in any other book. It all goes to show how my own understanding is constantly evolving and how, in writing my books, I am not only trying to communicate what I know but also struggling to figure things out to my own satisfaction.

In addition to the significance of the variety in the world and the darkness of the sky at night, I also discuss how the complexity of the world tells us not only that God plays dice with the Universe – an idea that Einstein abhorred – but that if He did not, there would be no Universe at all. I also discuss how direction of time – the reason why you grow old rather than young – appears to have been set when gravity 'switched on' about 380,000 years after the Big Bang, a discovery made by Larry Schulman while I was writing this book. And I describe Stephen Hawking's discovery, also made while I was writing this book, that the fact we live in a non-quantum world in which people never walk through two doors simultaneously implies that the Universe must have undergone a burst of super-fast expansion in the past. This is surely one of the most astonishing deductions to be made from everyday reality and underlines Hawking's unique genius. And there is more. But this is already too long for an introduction. I hope you enjoy my book.

Marcus Chown
London
February 2009

PART ONE
What the Everyday World Is Telling You about Atoms

The Face in the Window

*How, when you stand in front of a window, the most
shocking discovery in the history of science – that
ultimately things happen for no reason – is literally
staring you in the face*

'*Une difficulté est une lumière. Une difficulté insurmontable est un
soleil.*' (A difficulty is a light. An insurmountable difficulty is a
sun.)

Paul Valéry

'No progress without paradox.'
John Wheeler, 1985

*It is night-time and it is raining. You are staring dreamily out
of a window at the lights of the city. You can see the cars driving
past on the street and you can see the faint reflection of your
face among the runnels of water streaming down the pane.
Believe it or not, this simple observation is telling you something profound and shocking about fundamental reality. It is
telling you that the Universe, at its deepest level, is founded on
randomness and unpredictability, the capricious roll of a dice –
that, ultimately, things happen for no reason at all.*

The reason you can see the lights of the city outside and
simultaneously the faint image of your face staring back at
you is because about 95 per cent of the light striking the window goes straight through while about 5 per cent is reflected.
This is easy to understand if light is a wave, like a ripple on
water, which is the commonly held view. Imagine a speedboat

streaking across a lake and creating a bow wave which runs into a piece of partially submerged driftwood. Most of the wave just keeps on going, unaffected by the obstacle, while a small portion doubles back on itself. Similarly, when a light wave encounters the obstacle of a window, most of the wave is transmitted, while a small portion is reflected.

This explanation of why you see your face in a window is straightforward. It certainly does not appear to have any profound implications for the nature of ultimate reality. However, this is an illusion. Light is not what it seems. It has a trick up its sleeve which undermines this simple picture and changes everything. In the twentieth century, a number of phenomena were discovered that revealed that light behaved not as a wave, like a ripple spreading on a pond, but as a stream of bullet-like particles. For instance, there was the Compton effect, which revealed something very peculiar about the way light bounced, or 'scattered', off an electron. Discovered in 1897 by Cambridge physicist 'J. J.' Thomson, the electron was a particle smaller than an atom. In fact, it was one of its key constituents.

In 1920, the American physicist Arthur Compton decided to investigate what happened to light when it was shone on electrons. He had a picture in his mind of light waves bouncing off an electron like water waves off a buoy. If you have seen such a thing, you will know that the size, or 'wavelength', of the waves remains unchanged. In other words, the distance between successive wave crests is the same for the outgoing wave as the incoming wave. But in Compton's experiment this was not the case at all. After the light waves had bounced off electrons, their wavelength was *bigger* than before. And the more the direction of the light was changed in the encounter, the bigger the change in wavelength. It was as if the mere act of bouncing off an electron magically changed blue light, which is characterised by a short wavelength, into red light, which has a longer wavelength.[1] A longer, more sluggish wave turns out to be less energetic than a short, frenetic wave. So what Compton's experiments were telling him was that, when light bounced off an electron, it was somehow sapped of energy.

Compton's mental picture of what was going on was knocked for six. The light in his experiments was not behaving anything like a water wave bouncing off a buoy. In fact, the more he thought about it, the more he realised that it was behaving like a billiard ball hitting another billiard ball. When a ball is struck by the cue ball, it shoots off, carrying with it some of the energy of the cue ball. Inevitably, the cue ball loses energy. Electrons were known to be like tiny billiard balls, but light was known to ripple though space like a wave. Compton's experiments were unequivocal, however. Despite centuries of evidence to the contrary, light must also consist of particles like tiny billiard balls. For his groundbreaking work in confirming the particle-like nature of light, Compton was awarded the 1927 Nobel Prize for Physics.

More evidence that light behaved like a stream of particles came from the photoelectric effect, familiar to everyone who sees supermarket doors part like the Red Sea when they walk towards them. What triggers the doors to swish aside is the breaking of a beam of light by an approaching leg or a foot. The beam illuminates a 'photocell', a device containing a metal which spits out electrons whenever light falls on it. This happens because the electrons are only loosely bound to their parent atoms, so the energy delivered by the light is sufficient to kick them free. When someone breaks the light beam, the photocell is cast into shadow and the sputtering of electrons stops. The electronics are rigged in such a way that the instant the flow of electrons chokes off the doors open.

So what has the photoelectric effect got to do with the particle nature of light? If light is a wave, it is nigh on impossible to explain how it can deliver energy efficiently to a tiny, localised electron. Being spread out, a typical light wave will interact with a large number of electrons spread over the surface of the metal. Inevitably, some will get kicked out after others. In fact, calculations show that some electrons will be kicked out up to ten minutes after others. Imagine if the flow of electrons took ten minutes to build up in the photocell, so supermarket customers had to wait ten minutes for an automatic door to open.

Everything makes sense if the light is made of tiny particles and each interacts with a single electron in the metal. Rather than spreading its energy over large numbers of electrons, the light tied up in such 'photons' packs a real punch. Not only does each photon eject a single electron but it ejects it *promptly*, not after a ten-minute delay. Thank the particle-like nature of light for your prompt admission to a supermarket.

It was for explaining the photoelectric effect in terms of tiny chunks, or 'quanta', of light that Einstein won the 1921 Nobel Prize for Physics. Many people find this surprising. They wonder why he did not win the prize for 'relativity', the theory for which he is most famous and which changed for ever our view of space and time. Einstein himself, however, always saw relativity as a natural and unsurprising outgrowth of nineteenth-century physics.[2] He considered 'quanta', alone among his achievements, the only truly revolutionary idea of his life.

Einstein published his paper on the existence of quanta in

the same 'miraculous year' as his theory of relativity. Five years earlier, in 1900, the German physicist Max Planck had found a way to explain the puzzling character of the heat coming from a furnace by suggesting that atoms can vibrate only at certain permissible energies and that those energies come in multiples of some basic chunk, or quantum, of energy. Planck believed these quanta to be no more than a mathematical sleight of hand, with no physical significance whatsoever. Einstein was the first person to view them as truly real – as flying through space as a stream of photons in a beam of light.

The Matchbox that Ate a 40-Tonne Truck

Actually, the fact that light must in some circumstances behave as tiny, localised particles is forced on us by the most familiar of everyday phenomena – the emission of light by the filament of a light bulb and the absorption of light by your eye. The reason has to do with the make-up of the filament and your retina. Like all matter, they are made of atoms.

The idea that everything is made of atoms comes from the Greek philosopher Democritus, who, around 440 BC, picked up a rock or a branch or maybe it was a piece of pottery and asked himself: 'If I cut this object in half, then cut the halves in half, can I go on subdividing it like this for ever?' Democritus answered his own question. It was inconceivable to him that matter could be subdivided in this way for ever. Sooner or later, he reasoned, you must come to a tiny grain of matter which could not be cut in half any more. Since the Greek for 'uncuttable' was *a-tomos*, Democritus's ultimate grains of matter have come to be known as 'atoms'.

Democritus actually went further and postulated that

atoms come in a handful of different types, like microscopic Lego bricks, and that, by assembling them in different ways, it is possible to make a rose or a cloud or a shining star. But the key idea is that reality is ultimately grainy, composed of tiny, hard bullets of matter. It is an idea that has certainly stood the test of time.[3]

Atoms turn out to be very small. It takes more than a million to span a pinhead. Confirming their existence was therefore very hard. A lot of indirect evidence was accumulated in the age of science. However, remarkably, no one actually 'saw' an atom until 1980, when two physicists at IBM built an ingenious device called the Scanning Tunnelling Microscope.

The STM earned Gerd Binnig and Heinrich Rohrer the 1986 Nobel Prize for Physics. Basically, the device drags a microscopic 'finger' across the surface of a material, sensing the up-and-down motion as it passes over the atoms in much the same way that a blind person senses the undulations of someone's face with their finger. And, in the same way a blind person builds up a mental picture of the face they are feeling, the STM builds up a picture on a computer display of the atomic landscape over which it is travelling.

Using the STM, Binnig and Rohrer became the first people in history to look down, like gods, on the microscopic world of atoms. And what they saw, swimming into view on their computer screen, was exactly what Democritus had imagined 2,500 years earlier. Atoms looked like tiny tennis balls. They looked like apples stacked in boxes. Never, in the history of science, had someone made a prediction so far in advance of its experimental confirmation. If only Binnig and Rohrer had a time machine. They could have transported Democritus to their Zürich lab, stood him in front of their remarkable image and said: 'Look. You got it right.' Just like artists who die in obscurity, never having seen their reputations go stratospheric and their paintings sell for tens of millions of pounds, scientists may never live to see the spectacular success of their ideas.

Atoms, it turns out, are not the ultimate grains of matter. They are made of smaller things. Nevertheless, Democritus's idea that matter is ultimately grainy, not continuous, persists, with 'quarks' and 'leptons' now wearing the mantle of nature's uncuttable grains. But quarks, it turns out, are not important when it comes to the meeting of light and matter in your eye or in the filament of a light bulb. When light is absorbed or spat out, it is atoms that do the absorbing and spitting. And herein lies the problem.

An atom, according to our theory of matter, is a tiny, localised thing like a microscopic billiard ball. Light, on the other hand, is a spread-out thing like a ripple on a pond. Take visible light. A convenient measure of its size is its wavelength – the distance it travels during a complete up-and-down oscillation, or double the separation of successive wave crests. The wavelength of visible light is about 5,000 times bigger than an atom. Imagine you have a matchbox. You open it and out drives a 40-tonne truck. Or say a 40-tonne truck is driving towards you, you open your matchbox and the truck disappears inside. Ridiculous? But this is precisely the paradox that exists at the interface where light meets matter.

How does an atom in your eye swallow something 5,000 times bigger than itself? How does an atom in the filament of a light bulb cough out something 5,000 times more spread out? The British survival expert Ray Mears said during one of his TV programmes: 'Nothing fits inside a snake like another snake.' Apply this logic to the interface between light and matter. If light is to fit inside an atom, which is small and localised, it too must be small and localised. The trouble is there are a thousand instances – most notably Young's double-slit experiment – where light shows itself to be a spread-out wave.

In the first decades of the twentieth century, physicists too went round and round in circles, trying desperately to resolve paradoxes of this kind. As the German physicist Werner Heisenberg wrote: 'I remember discussions which went through many hours until very late at night and ended almost in despair; and when at the end of the discussion I went alone for a walk in the neighbouring park I repeated to myself again and again the question: Can nature possibly be so absurd as it seemed to us in these atomic experiments?'

A paradox where one theory predicts one thing in a particular circumstance and another theory something quite different is often hugely fruitful. It tells us that one theory at least is wrong. And the bigger and more well-established the theories which are at loggerheads, the more revolutionary the consequences. In the case of light being emitted from a light bulb or being absorbed by your eye, the two theories which predict conflicting things are the wave theory of light and the atomic theory of matter. And they are two of the biggest and most well-established theories of all.

So which theory is wrong? The extraordinary answer embraced by physicists is both. Or neither. Light is both a wave and a particle. Or, rather, it is something for which we have no word in our vocabulary and nothing we can compare it with in the everyday world. It is fundamentally ungraspable – like a three-dimensional object is to creatures confined to the two-dimensional world of a sheet of paper, with no concept of up above or down below. All they can ever experience are 'shadows' of the object, never the object in its entirety. Similarly, light is not a wave or a particle but 'something else' that we can never grasp completely. All we can see are its shadows – in some circumstances its wave-like face and in others its particle-like face.

Clearly, atoms do spit out light. But, just as clearly, visible light is many thousands of times bigger than an atom that spits it out. Both facts are incontrovertible. The only way to resolve the paradox, therefore, is to accept something that sounds like sheer madness – that light is both thousands of times bigger than *and* smaller than an atom. It is both spread-out *and* localised. It is both a wave *and* a particle. When it travels through space, light travels like a ripple on a pond. However, when it is absorbed or spat out by an atom, it behaves like a stream of tiny machine-gun bullets. Imagine you are standing by a fire hydrant in New York's Times Square and simultaneously spread out like a fog throughout Manhattan. Ridiculous? Yes. Nevertheless, that is the way light is.

The wave picture of light was correct. So too was the particle picture. Paradoxically, light is both a wave and a particle.

A World that Defies Common Sense

Should we be surprised to find that light is fundamentally different from anything in the everyday world? Should we be surprised that it is ungraspable in its entirety, that its properties are counter-intuitive, that they defy common sense? Perhaps it helps to spell out what we mean by intuition or common sense. Really, it is just the body of information we have gathered about how the world around us works. In evolutionary terms we needed that information to survive on an African plain in the midst of a lot of creatures which were bigger, faster and fiercer than us. Survival depended on having vision that enabled us to see relatively big objects between us and the horizon, hearing that enabled us to hear relatively loud sounds, and so on. There was no survival value in developing senses that could take us beyond the world of our immediate surroundings – eyes, for instance, that could show us the microscopic realm of atoms. Consequently, we developed no intuition whatsoever about these domains. We should, therefore, not be surprised that when we began to explore the domain of the very small compared to our everyday world, we found counter-intuitive things. An atom is about 10 billion times smaller than a human being. It would be surprising if it behaved in any way like a football or a chair or a table, or anything else in the world of our senses.

The first person to realise that the fundamental reality that underpins the everyday world is totally unlike the everyday world was the Scottish physicist James Clerk Maxwell, arguably the most important physicist between the time of Newton and Einstein (tragically, he died of stomach cancer, aged only 48). His great triumph, in the 1860s, was to distil all magnetic and electrical phenomena into one neat set of formulae. 'Maxwell's equations' are so super-compact you could write them on the back of a stamp (if you have small hand-writing!).

Up until the time of Maxwell, physicists modelled the world in terms of things they could see around them. They talked, for instance, of a Newtonian 'clockwork' Universe. Maxwell was no different. Initially, when struggling to understand how a magnet reached out and tugged on a piece of metal, for instance, he imagined the space between the magnet and the metal filled with invisible toothed cogs. A cog pressed tight up against the magnet turned another cog, which turned another, and so on. In this way, the force was transmitted from the magnet to the metal. When the picture did not fit his observations of magnetism, Maxwell modified it, imagining that the cogs were made of springy material that flexed as they turned. When this did not work either, he threw up his hands in despair and dispensed with such 'mechanical' models. Nature, he realised, was not like anything in everyday experience.

Instead of invisible cogs turning, Maxwell imagined ghostly electric and magnetic 'force fields' permeating space, with no parallel in the everyday world. It was a seismic break with the past. In the long term it would liberate physics, enabling Einstein to imagine gravity as a warpage of four-dimensional space–time and present-day physicists to hypothesise that the fundamental building blocks of matter are tiny strings of mass-energy vibrating in an unimaginable space of ten dimensions.

It took a while for physicists to learn the hard lesson that, in their quest to understand fundamental reality, they would have to do without the safety net of everyday intuition. They had still not learnt the lesson, in fact, when in the first decades of the twentieth century the titanic collision between the theories of light and matter spawned the wave–particle theory of light.

God Plays Dice

If light behaves as a stream of particles – and this is the point of this discussion – it has serious implications for understanding why you can see the reflection of your face in a window. Why? Well, what is perfectly straightforward to explain if light is a wave – remember the wave from the speedboat hitting the partially submerged wood and being partially reflected – is fiendishly difficult to explain if light is instead a stream of bullet-like particles. Photons, after all, are identical. However, if they are identical, surely they should be affected identically by a pane of glass. Either they should all be transmitted or they should all be reflected. So how can 95 per cent go through and 5 per cent bounce back?

This is a classic case of a physical paradox – a situation in which one theory, the particle theory of light, predicts one thing, whereas our common-sense experience tells us something contradictory. Our experience is clearly trustworthy – we can indeed see the scene outside a window and simultaneously the faint reflection of our face. Consequently, something must be awry with our idea of photons.

There is only one logical possibility: each photon must have a 95 per cent *chance* of being transmitted and a 5 per cent *chance* of being reflected. It may seem an innocuous fact, but actually it is a bombshell dropped into the heart of physics. For if we can know only the chance, or 'probability', of a photon going through a window or coming back, then we have tacitly given up all hope of knowing for sure what an individual photon will actually do. As realised by Einstein – ironically, the first person to propose the existence of the photon – this was a catastrophe for physics. It was utterly incompatible with everything that had gone before. Physics was a recipe for predicting the future with total confidence. If, at midnight, the Moon is over here in the sky, using Newton's law of gravity we can predict that at the same time tomorrow night it will be over there – with 100 per cent certainty. But take a photon impinging on a window pane. We can never predict with certainty what it will do. Whether it is transmitted or reflected is totally, utterly random, determined solely by the vagaries of chance.

This kind of chance is not the type familiar from the roll of a dice and the spin of a roulette wheel. It is far more fundamental – and sinister. If all the myriad forces acting on a dice were known, a physicist with a big enough computer and enough dogged patience could use Newton's laws of motion to predict the outcome. The problem is there are so many factors influencing the trajectory of a dice – from the initial impetus given it by a gambler to the currents of air that buffet it to the roughness of the tabletop over which it tumbles – that it is beyond anyone's capabilities to pin down all of them with the necessary precision to predict the out-come with certainty.

But the key thing to recognise is that our ignorance of the factors influencing the roll of a dice is merely a practical problem. It is not impossible that, in the future, someone with sufficient tenacity – not to mention time on their hands – might be able to determine to the required degree of accuracy all the forces acting on a dice. The point is, the roll of a dice is not inherently unpredictable. It is only unpredictable in practice.

Contrast this with a photon. What a photon does when it encounters a pane of glass is utterly unpredictable – not merely in practice but in principle. It is not a matter of us being ignorant of all the factors that influence what it does. There are no factors to be ignorant of. A photon goes through a window rather than bouncing back out of sheer bloody-mindedness – for no reason at all.

In the day-to-day world every event is triggered by a prior event. A cause always precedes an effect. The dice comes up

the number it does because of the effect of all the forces acting on it. You trip and stumble while out walking because a
paving stone is loose and catches the heel of your shoe. But
what a photon does on encountering a window pane is triggered by no prior event. It is an effect without a cause.
Though the probability of a dice coming up 'six' can be
determined in principle, there is no prior event from which
the probability of a photon going through a window can be
determined, no hidden machinery whirring beneath the skin
of reality. It is nature's bedrock, its bottom line. There is
nothing deeper. For some mysterious reason, the Universe is
simply constructed this way.[4]

The kind of unpredictability that characterises photons at
a window pane in fact characterises their behaviour in all
conceivable circumstances. It is, actually, typical of the
behaviour of not just photons but all denizens of the microscopic world of atoms and their constituents – the ultimate
building blocks of reality. An atom of radium can disintegrate, or 'decay', its central 'nucleus' exploding violently like a
tiny grenade. But there is absolutely no possibility of predicting exactly when an individual radium nucleus will self-
destruct, only the probability that it will happen within a
particular interval of time.

The unpredictability of the microscopic world is unlike
anything human beings have ever come across before. It is
something entirely new under the sun. This is why Einstein
got the Nobel Prize for deducing the particle-like nature of
light from the photoelectric effect, and not for the theory of
relativity. He – and the Nobel committee – realised it was a
truly revolutionary discovery.

The recognition that the microscopic world is ultimately
controlled by irreducible, random chance is probably the

single most shocking discovery in the history of science. Ironically, it so appalled Einstein that he famously declared: 'God does not play dice with the universe.' (The great quantum pioneer Niels Bohr retorted: 'Stop telling God what to do with his dice.') He steadfastly refused to believe that things at a fundamental level in the Universe happened for no reason at all. The bitter irony, not lost on Einstein, was that he was the one who, by postulating the existence of the photon, had inadvertently set loose the genie of randomness in the heart of physics.[5]

To Einstein's dismay, other physicists in the 1920s appeared to embrace the quantum idea that things can happen for no reason at all. But Einstein's intuition told him something important. If naked randomness was admitted into the heart of the world, it would inevitably spawn even more shocking consequences – consequences so troubling, he believed, that physicists would be forced to abandon the whole quantum idea. It took until 1935 but, eventually, Einstein found what he was looking for. Working with two other physicists – Nathan Rosen and Boris Podolsky – he discovered that if quantum theory was right, then an inescapable consequence was that two atoms could influence each other instantaneously, even on opposite sides of the Universe.

To appreciate how Einstein came to such a conclusion requires a digression. This chapter began with the assertion that the reflection of your face in a window pane is easy to understand if light is a wave like a ripple on a pond. But there was no mention of how we ever came to suspect that light is a wave. After all, it does not *look* like a wave.

Light Is a Wave

The man who demonstrated that light was a wave was the Englishman Thomas Young. He was a polymath who not only made the first breakthrough in deciphering the Egyptian hieroglyphs on the Rosetta Stone but realised that the eye must contain separate receptors for blue, green and red light. Arguably his greatest achievement, however, was to lay bare the wave nature of light.

Young had a strong suspicion that light was a wave rather than a stream of bullet-like 'corpuscles', as Newton believed. In 1678, the Dutch scientist Christian Huygens had found that if light were a wave rippling through space, it was possible to explain many optical phenomena, such as the reflection of light by a mirror and the bending, or 'refraction', of the path of light by a dense medium such as glass. Huygens' wave theory even predicted the correct bending of light as it travelled from air into a block of glass, whereas Newton's did not – at least not without some tinkering. Such was Newton's God-like standing, however, that Huygens' theory was pretty much ignored – until Young.

A central characteristic of waves of any kind is that, when they pass through each other, they alternately reinforce and cancel each other out. They reinforce, or 'constructively interfere', where the peaks of one wave coincide with the peaks of another; and they cancel, or 'destructively interfere', where the peaks of one coincide with the troughs of another. This 'interference' is hypnotic to watch in a puddle during a rain shower. As the concentric ripples from impacting raindrops spread through each other, they alternately boost and nullify each other.

Young knew of this effect. He knew also that if a similar

effect occurred with light, the fact it was not visible to the naked eye could only mean that the crests of light waves must be separated by far less than the width of a human hair, one of the smallest things discernible to the human eye. Making the interference of such tiny waves visible was a challenge, to say the least. But Young rose to it.

The key, he realised, was to create two similar sources of concentric ripples just like those spreading from two raindrops that puncture the skin of a pond. As the ripples spread through each other, they would interfere. At the places where the ripples destructively interfered, cancelling each other out, there would be darkness; and at the places where they constructively interfered, bolstering each other, there would be enhanced brightness. The dark and light regions would alternate. To see them it would be necessary only to put some kind of white screen at a location where the concentric ripples overlapped. This would reveal the interference as a pattern of alternating light and dark zebra stripes, not unlike a supermarket bar code.

It was crucial for the success of Young's experiment that the light be of a single colour, or as close to a single colour as was possible. Different colours of light are today known to correspond to different wave sizes, or 'wavelengths', with the crests of red light being roughly twice as far apart as the crests of blue. Young may have suspected this. Since demonstrating the interference of light required perfect cancellation and perfect reinforcement of the overlapping light waves, it could happen only if there was light of a single colour.

In 1801, Young created his two sources of concentric ripples by shining light on one side of an opaque screen with two closely spaced, parallel slits cut in it. On the other side of the screen, the light emerged from each slit, spreading out and passing through the light from the other slit. In the region where the ripples overlapped Young interposed his white screen. And there, triumphantly, he saw a pattern of light and dark stripes – the unmistakable signature of interference. Beyond any doubt light was a wave. The reason it was not obvious to the naked eye was because the waves were so small: only a thousandth of a millimetre from crest to crest.[6]

Why is it necessary to know about an experiment at the beginning of the nineteenth century that demonstrated the wave nature of light? Because this was not the end of the story for Young's double-slit experiment. Not by a long chalk. In the twentieth century, it reappeared in a new incarnation. And, remarkably, this time it demonstrated not the wave character of light but something else – something scarcely believable. That it is possible for a single microscopic entity – a photon or an atom – to be in *two places at once*.

Waves Inform Particles

Recall that Young shone light of a single colour, or wavelength, onto an opaque screen into which were incised two closely spaced, parallel slits. Each slit acted as a source of secondary light waves, just as two stones dropped in a pond together act as sources of concentric ripples. And, just as the ripples from two stones pass through each other, alternately reinforcing and cancelling, so too do the light ripples from the two slits. Where they reinforce, the light is boosted in brightness; where they cancel, it is snuffed out, leaving darkness. Young interposed a second screen in the region where the waves overlap. And there for all to see were alternating bands of light and dark. Beyond a shadow of a doubt, light was a wave.

But, beyond a shadow of a doubt, it was also a stream of

particles. Arthur Compton had shown it to bounce off electrons as if it was made of tiny billiard balls, and there was also the photoelectric effect, in which individual particles of light liberated individual electrons from the surface of a metal. The key question therefore was: how is it possible to reconcile this with Young's experiment?

Think about photons of visible light. Each carries very little energy. This is why nobody noticed their existence before Einstein. If photons carried large amounts of energy, when someone used a dimmer switch to turn up a light, the brightness would jump in abrupt steps from zero to some minimum brightness, then double that brightness, triple that brightness, and so on. We never see a light source brighten like this. And the reason is that individual photons carry so little energy that the steps, though present, are simply too minuscule to be discernible with the naked eye.

The light source in Young's experiment is also composed of trillions upon trillions of tiny photons. Although this explains why its particle nature is not obvious, it does not explain how the photons conspire to form an interference pattern of dark and light bands, the unequivocal signature of waves, not particles. One possibility is that when large numbers of photons are present, their particle-like nature is somehow washed out in favour of their wave-like nature, that they lose their individuality like a lone person in a crowd at a football match. But what if we force light to show its particle hand? This can be done by carrying out Young's experiment with a source of light so weak that it contains not trillions upon trillions of photons but only a few. If the source is so weak that photons arrive at the slit in the screen one at a time, with long intervals between, there will be no doubt at all that we are dealing with particles.

The human eye cannot detect single photons, so the arrival of photons on the second screen will be invisible. Nevertheless, this can be overcome by covering the screen with an array of sensitive detectors capable of registering individual particles of light. Think of them as tiny buckets which collect photons, just as real buckets collect raindrops. If the photon buckets are connected to a computer, what they pick up can be displayed on a screen and so made visible to the human eye.

If we set up this high-tech version of Young's experiment, what might we expect to see? Well, it is a fundamental feature of interference that it takes two waves to mingle, or interfere, with each other. In the case of Young's experiment, the two sets of waves emerge like concentric ripples from the two slits in the opaque screen. However, if photons are arriving at the screen one at a time, with large gaps of time in between, then it stands to reason there will only ever be one photon at a time emerging from one slit or the other. Such a solitary photon will have no other photon to mingle with. There can be no interference. So, after the experiment has been running a long while and lots of photons have gone through the two slits and peppered the second screen, the pattern on the computer monitor should simply reveal two parallel, bright bands – the images of the two slits.

But this is not what happens.

At first, the computer screen appears to show the photons raining down all over the second screen, as if fired from some kind of scattergun. However, as the experiment continues, something remarkable happens. Slowly but surely a pattern begins to emerge, like Lawrence of Arabia appearing out of the desert dust, built up a photon at a time from the particles intercepted by the tiny light buckets. And it is not just any pattern. It is a pattern of alternating light and dark bands, precisely the parallel interference stripes seen by Young in 1801. But how can this be? Interference arises from the mingling of waves from two sources. Here, the light is so weak that it is demonstrably made of particles – the light-bucket detectors, after all, register them one click at a time – and each photon has no other with which to mingle.

Welcome to the weird world of the quantum. Photons doing things for absolutely no reason at all turns out to be merely the beginning of the madness.

It seems that photons, even when there are so few of them that they are undeniably individual particles, have some awareness of their wave nature. After all, they end up on the second screen at exactly the places that waves emerging from the two slits would reinforce each other, while studiously avoiding the places where waves from the two slits would cancel. It is as if there is a wave associated with each photon that somehow directs it where to go on the screen.

And this is pretty much the picture most physicists, rightly or wrongly, carry in their minds. There is a wave associated with a photon. It informs it where to go or what to do. There is a twist, however. The wave is not a real, physical wave that can be seen or touched like a wave on water. Instead, it is an abstract, mathematical thing. Physicists imagine this quantum wave, often called the 'wave function', as extending throughout space. Where the wave is big, or highly peaked, there is a high chance, or probability, of finding the photon; and where it is small, or relatively flat, there is a low probability of finding it. To be a little more specific, the chance, or probability, of finding a particle at a particular location in space is the square of the height of the quantum wave at that location. Quantum waves can mingle and interfere and, when they do, the interference pattern produced determines where the photons are most likely to be found.

It is a hard picture to get your head around. Nevertheless, it hints at a profound duality in nature. Not only can light waves behave as particles – photons – but photons in turn can behave like waves, albeit abstract quantum waves.

As already pointed out, the consequence of light waves behaving as particles is pretty earth-shattering. The world of photons – and everything else – is ultimately orchestrated by random chance. And it turns out that the consequence of photons behaving as waves is equally earth-shattering. A single photon can be in two places at once (or do two things at once), the equivalent of you being in London and Paris at the same time. How come? Well, if photons can behave like waves, then it follows they can do all the things that waves can do. And there is one thing that waves can do which, although it has mundane consequences in the everyday world, has remarkable consequences in the microscopic world.

Two Places at Once

Imagine the sea on a stormy day. Big rolling waves driven by the wind are marching across the surface. Now imagine the sea a day later, when the storm has passed. The surface of the water is calm except for tiny ripples, ruffles caused by the light breeze. Now it is also possible to have big rolling waves with tiny wind-ruffled ripples superimposed. And this, it turns out, is a general feature of waves of all kinds. If two different waves are possible, a combination of those two waves is always possible. In the case of ocean waves, this has a consequence hardly worthy of note. But in the case of the quantum waves associated with photons, which inform them where to be and what to do, the consequences are pretty amazing.

Imagine a quantum wave which is highly peaked on one side of a window pane, so there is a high probability of finding it on that side. Now imagine another quantum wave which is highly peaked on the other side. Nothing untoward here. However, since both waves are individually possible, a wave that is a combination, or 'superposition', of both is also possible. In fact, it is required to exist. But this corresponds to a photon that is on both sides of the window pane at the same time. A photon that is simultaneously transmitted and reflected. Surely this is impossible?

Think back to Young's double-slit experiment again. Recall that, to create an interference pattern, two things must mingle. One way to think about this is from the wave perspective. In this case, the quantum wave associated with each photon spreads out in concentric ripples from the slits in the opaque screen. But the other way to think of it is from the particle point of view. In this case, each photon arriving at the opaque screen is in two places at once. This enables it to go through both slits simultaneously and mingle with itself.

The ability of a photon to do two things at once is a direct consequence of the fact that if two waves are possible, a combination of those two waves is also possible. But nature does not stop at just two waves. If any number of waves are possible – three, 99 or 6 million – a combination of all of them is also possible. A photon can not only do two things at once, it can do many things at once.

It turns out there is an equation – a recipe, if you like – which predicts precisely how the quantum wave corresponding to a photon, or anything else, spreads through space. It was devised by the Austrian physicist Erwin Schrödinger, and his equation answers a quantum conundrum, namely, if the Universe is fundamentally unpredictable, at the mercy of the quantum roll of a dice, how is it that the everyday world is largely predictable? How is it that we can predict with almost complete certainty that if you get caught out in the rain, you will get wet? Or that the Sun will rise tomorrow morning?

The Schrödinger equation shows that what nature takes away with one hand it grudgingly gives back with the other. Yes, the Universe is fundamentally unpredictable. However – and this is the key thing – the unpredictability is predictable. We cannot know for certain what a photon or any other microscopic particle can do. But, with the aid of the Schrödinger equation, we can know the probability of it doing one thing, the probability of it doing another, and so on. And this, it turns out, is enough to ensure that we live in a largely predictable world.

More than that. Quantum theory is the most successful physical theory ever devised. Its predictions match what we see in experiments to an obscene number of decimal places. Quantum theory has literally made the modern world possible, not only giving us lasers and computers and iPods but also an understanding of why the Sun shines and why the ground beneath our feet is solid. It is ironic that we have this hugely successful theory which, on the one hand, is a remarkable recipe for building things and understanding our world, yet, on the other, provides a window onto an Alice-in-Wonderland world that is stranger than anything we could possibly have invented.

Instantaneous Influence

But if you think a photon doing something for absolutely no reason at all or being in two places at once is bad, there is worse to come. And this is where Einstein, Rosen and Podolsky came in. They highlighted a consequence of quantum theory they believed was so ridiculous that it must force all reasonable people to drop it. Think of the particle nature of light waves, which leads to naked unpredictability, and the wave nature of photons, which enables a photon to be in two places at once. Now imagine combining them. The result, Einstein's team discovered, is a new, even weirder phenomenon: instantaneous communication between separated locations of space, even if those locations are on opposite sides of the Universe.

Actually, a third ingredient is required to conjure up the new phenomenon. But that ingredient is something so fundamental that it transcends quantum theory. It is a conservation law. Physicists have discovered a number of these. For instance, there is the law of conservation of energy. This states that energy can never be created or destroyed, merely changed from one form into another. In a light bulb, for example, electrical energy is converted into light energy and heat energy. In your muscles, the chemical energy derived ultimately from your food is converted into the mechanical energy of movement of your muscles.

In 1918, one of the great unsung heroines of science, the German mathematician Emmy Noether, made a surprising discovery about conservation laws in physics. She discovered that they are merely consequences of deep 'symmetries' of nature – things that stay the same even when there is a change in our viewpoint. For instance, the conservation of energy stems from 'time translation symmetry': the fact that, if we do an experiment now or translated in time – say, next week or next year – all things being equal, we will get exactly the same result. Another deep symmetry of nature is 'rotational symmetry'. If we carry out an experiment with our equipment aligned north–south and rotate it to, say, the east–west direction, we will get the same result. The law which stems from this innocuous symmetry is the conservation of angular momentum, angular momentum being a quantity which is a measure of a rotating body's tendency to keep turning. The Earth, spinning on its axis, has a very large angular momentum, and so is likely to stay spinning for a long time.

It turns out that microscopic particles such as photons possess a quantum property called 'spin'. In common with irreducible randomness, it has no analogue whatsoever in the everyday world. As far as we know, photons as they fly through space are not actually spinning like the Earth spins on its axis. Their spin is 'intrinsic'. Nevertheless, they *behave* as if they are spinning. Specifically, a photon has two possibilities open to it: it can behave as if it is corkscrewing in a clockwise manner about its direction of motion at a particular spin rate; or it can behave as if it is corkscrewing in an anticlockwise manner at the same rate.

The key thing is that quantum spin obeys the law of conservation of angular momentum. And the law, applied to photons, says that if two photons are created together, their total spin can never change. Say they are born together and one is spinning clockwise and the other anticlockwise. Their spins cancel each other out. In the jargon, physicists say their total spin is zero. In this case, the conservation of angular momentum requires that the total spin of photons must remain zero for ever, or until some process destroys them.

Nothing peculiar or controversial about this.

But consider a real process that creates two oppositely spinning photons. The electron, the tiny particle that orbits inside atoms, has an 'antiparticle' twin called the positron. It is a characteristic of all particles and such 'antimatter' twins that when they meet, they destroy, or 'annihilate', each other. Now an electron and a positron have an intrinsic spin just like a photon. It has a different magnitude to that of a photon, but that is not important here. The important thing is that just before they annihilate, the electron and positron are spinning in opposite directions, so their total spins cancel each other out. This ensures that the two photons created must also have spins that cancel out. One must be spinning clockwise and the other anticlockwise.

But here comes the quantum twist. The conservation of angular momentum requires only that the spins of the two photons that fly away from the annihilation are *opposite to each other*. But there are two possible ways this can happen. The first photon can be spinning clockwise and the second anticlockwise. Or the first photon can be spinning anticlockwise and the second clockwise. Remember, however, this is a quantum world. Each possibility is represented by a quantum wave. And, if two waves are possible, recall that a combination is also possible – required, in fact.

So as the newborn photons fly off – and they fly off in opposite directions – they exist in a weird quantum 'superposition'. Just as a single photon can be on both sides of a window pane at the same time, the two photons are simultaneously spinning clockwise-anticlockwise and anticlockwise-clockwise. Maybe you do not see the bombshell lurking here. Don't worry. Nobody did. It took Einstein to see it.

In addition to using the conservation of angular momentum, we have so far used one quantum ingredient – the quantum superposition. That leaves only the second quantum ingredient – unpredictability. Say we have arranged for there to be a detector which will intercept the first photon and determine its spin. Now it is impossible to predict for certain which way the photon will be spinning – even in principle. The quantum world is characterised by irreducible randomness. All we can know is that there is a 50 per cent chance that when we detect the photon, we will find it spinning clockwise, and a 50 per cent chance we will find it spinning anticlockwise.

Say we detect the first photon and find it is spinning clockwise. Now here comes the bombshell. Instantaneously, the second photon must start spinning anticlockwise. The photons were born spinning in opposite directions, after all, and the conservation of angular momentum requires that they must always spin in opposite directions. If, on the other hand, we detect the first photon and find that it is spinning anticlockwise, the second photon must start spinning clockwise instantaneously. What is mind-blowing about this is that there is no reference whatsoever to how far apart the photons are. If one photon is found to be spinning one way, its twin must react *instantaneously* so as to ensure that it is spinning in the opposite sense – even if the photons are on opposite sides of the Universe.

Quantum theory, as spectacularly shown by Einstein, Rosen and Podolsky, permits the insanity of instantaneous influence at a distance. It implies that particles born together for ever after behave as if, in some sense, they are a single joined-at-the-hip particle rather than two separate ones. They *know* about each other. Their properties are inextricably entwined or, in the quantum jargon, 'entangled'. Instantaneous influence is synonymous with some kind of ghostly influence travelling between quantum particles at *infinite speed*. However, this flies in the face of Einstein's special theory of relativity, which maintains that no influence can travel faster than light – 300,000 kilometres per second.

Everything can be traced back to the interaction of three things: superposition, unpredictability and the conservation of angular momentum. Because two photons are in a super-position, the state of the two particles – whether they are spinning clockwise-anticlockwise or anticlockwise-clock-wise – is not determined for sure until the spin of one parti-cle is observed. But when it is measured, the outcome is unpredictable. Yet the conservation of angular momentum somehow operates to give the second particle knowledge of its partner's spin so that it can instantaneously adopt the opposite spin.

It is the subtle interplay of these three factors that predicts the existence of instantaneous influence, technically known as 'non-locality'. And actually the conservation of angular momentum is not essential. There is absolutely no reason why instantaneous influence could not be demonstrated by substituting another conservation law, like the conservation of energy, for the conservation of angular momentum. It would simply require a bit of ingenuity to concoct a situa-tion in which instantaneous influence was explicit.

Some popular books maintain that two entangled parti-cles are like a pair of gloves. Imagine taking one glove from your drawer without looking at it, packing it in a bag and driving a long way away before opening the bag and check-ing it. If you find it is a left-hand glove, you will of course know immediately that the glove left behind in your drawer is a right-hand glove, and vice versa. But this is to misunder-stand (and to trivialise) the magic of entanglement. Two sep-arated quantum particles are not like two gloves. In the former case, one glove fits a left hand and one fits a right hand, and this is true for all time, or at least while the two gloves are in existence. If the glove you have carried with you turns out to be a right-hand glove, it was a right-hand glove before you opened your bag, which means the stay-at-home glove was a left-hand glove. There is no need for any signal to travel to the stay-at-home glove to tell it to be a left-hand glove. It was a left-hand glove all along.

Contrast this with two photons. If each is like a glove, it is a weird kind of glove, one that is neither left- nor right-hand, or rather a glove which has no pre-existing property of left-ness or rightness. This property is determined only when you take it out of your bag and look at it, at which point it plumps, utterly randomly, for being a right-hand glove or a left-hand glove. And the left-behind glove, which also had no pre-existing property of leftness or rightness, must respond, instantaneously, by becoming the opposite of its partner. It is the fact that the glove (or photon) has no state – and then that state is determined totally randomly – that forces there to be some ghostly connection between it and its partner at the moment its state is determined.

With non-locality, Einstein was convinced he had finally come up with a ridiculous prediction, one so stark-staring bonkers that it must mean that quantum theory was not nature's final word. The trouble is, the ridiculous phenomenon predicted by Einstein has actually been observed – by a French physicist called Alain Aspect. In 1982, a quarter of a century after Einstein's death, Aspect showed that photons on one side of his laboratory at the University of Paris Sud responded to photons on the other side as if some ghostly influence had passed between them significantly faster than the speed of light. Einstein was wrong. Quantum theory had passed yet another stringent test. The reality it described might be ridiculous, it might be unpalatable, but that was tough. It was simply the way it was.

Being able to communicate at infinite speed, in total violation of Einstein's cosmic speed limit set by the speed of light, would be a wonderful thing. However – and wouldn't you just believe it – what nature gives with one hand – the tantalising possibility of instantaneous, *Star Trek*-like communication – it takes away with the other. It is all down to randomness again. The only information that can be sent using instantaneous influence is the spin state of a photon. But if the sender is to exploit non-locality, they must send each photon of the message in a superposition of spinning clockwise and anticlockwise. Perhaps clockwise can encode a 'o' and anticlockwise spin a '1'. But if each photon is in a superposition of states, it will only have a 50 per cent chance of being a 0 and a 50 per cent chance of being a 1. The only message that can be sent is a random sequence of 0s and 1s, as useless a message as a series of random coin tosses. Einstein's speed limit of the speed of light is not violated because it turns out it is an upper limit on the speed of 'information'. Nature imposes no speed limit on the transmission of unusable gibberish. And that is all non-locality, as amazing as it seems at first sight, permits.

We have come a long way from the reflection of your face in the window. The image staring back at you tells you that the microscopic world of photons must be orchestrated by random chance. But in taking into consideration the wavelike behaviour of photons – which implies they can do two things at once – we arrive at non-locality. Many physicists consider this instantaneous influence the greatest mystery of quantum theory. It is fair to say that nobody knows what it means for the Universe at large. However, there is one thing we know for sure. All the Universe's countless particles were born together 13.7 billion years ago in the fireball of the Big Bang. Consequently, the ghostly ties that bind two spinning photons must, in some sense we do not quite understand yet, bind you and me to the atoms in the most distant stars and galaxies.

2

Why Atoms Rock and Roll All Over the Place

How the fact you do not fall through the floor is telling you that something is stopping the microscopic constituents of matter from being crushed too small

'Atoms are completely impossible from the classical point of view.'
Richard Feynman

'And maybe it's because my atoms won't stand still
because they want to rock and roll all over the place –'
Adrian Mitchell ('The Sun Likes Me',
Heart on the Left)

The ground beneath your feet is solid. The book you are holding in your hands is solid. You are solid. You probably think these things are unremarkable. But they are not. You see, 99.9999999999999 per cent of matter is empty space. The ground you are standing on is more tenuous than the most tenuous morning mist. This book is no more than a ghost book, the words you are reading no more than ghost words. You – sorry to say – are a ghost. Of course, if the ground is so incredibly insubstantial, you might be wondering how it can possibly be supporting your weight. Why don't you fall through the ground the way you would if it really were a morning mist? The answer is that there is something preventing the microscopic building blocks of matter from being even remotely in the same neighbourhood as each other. A mysterious force so ferocious that it drives apart the electrons and nuclei of atoms, stiffening matter as if its constituent particles were bolstered by a network of invisible girders. It is the existence of this force that is ultimately why the ground under your feet – despite being so incredibly tenuous – can resist your weight.

To understand why matter is so overwhelmingly made up of emptiness it is first necessary to know about atoms. As already mentioned, the idea that everything is made of atoms is due to the Greek philosopher Democritus.[1] He reasoned not only that matter is ultimately made of tiny, indivisible grains, but that those grains come in only a limited number of different kinds. By arranging these microscopic Lego bricks in different ways, it is possible to make a tree or a table or a human being. Everything is in the combinations.

It is far from obvious that matter is grainy rather than continuous. And that, Democritus maintained, is because atoms are much too small to see or touch directly. But, two millennia after Democritus, scientists began to accumulate indirect evidence of atoms. For instance, they realised that the behaviour of a gas such as steam made sense if it were made of a multitude of tiny atoms, flying hither and thither like a swarm of angry bees. In 1660, the Irish scientist Robert Boyle discovered that if the volume of a box containing a gas is halved – by pushing in a movable wall, or 'piston' – it takes twice as much force as before to hold the piston against the pressure pushing back. If the volume is reduced to a third, it takes three times the force. And so on. This observation, known as Boyle's Law, makes perfect sense if the pressure exerted by the gas is simply the jittery force of countless atoms hitting the piston like rain on a tin roof. If the volume of the container is halved, the atoms have half the distance to travel between hitting one wall and hitting the opposite one. Consequently, they strike the walls twice as often, exerting twice the pressure. Similarly, if the volume of the container shrinks to a third, the atoms have a third as far to travel, so they bounce off the walls three times as often, tripling the pressure.

So much for the evidence that atoms are tiny grains in motion. What about the evidence that they come in different types? This was very hard to obtain since, according to Democritus, the very reason there is so much bewildering variety in the world is that atoms do not stick simply to their own kind but at every opportunity link up with other types of atom. However, as Einstein observed: 'Subtle is the Lord but malicious he is not.' It turns out that some materials, such as gold, cannot be broken into anything more basic, by heat, acid or any other means. Such 'elemental' materials give every indication of being vast agglomerations of a single type of atom.

The first person to identify such materials was the French aristocrat Antoine Lavoisier. His list, compiled in 1789, contained 23 'elements'. Actually, some turned out not to be elemental, but Lavoisier had set the ball rolling. While he was unable to add to his list on account of losing his head to the guillotine five years later, others did. The problem was that, by the middle of the nineteenth century, the number of known elements had climbed to more than 50 – rather more than the handful of different atoms Democritus had envisioned as the building blocks of all matter. Today, we know of 92 naturally occurring elements, ranging from hydrogen, the lightest, to uranium, the heaviest. Why so many?

One possibility was that atoms were not the ultimate building blocks of matter but instead were made of smaller things. This proposal was made by a London chemist at the beginning of the nineteenth century. William Prout had compared the weights of different elements and discovered a striking pattern: most come in exact multiples of the weight of hydrogen, the lightest element.[2] A pattern, by definition, is the repetition of a simpler, more basic building block. Take the number 878787878787878787878787. The simpler building block from which it is made is the number '87'. Similarly, the pattern that Prout had found in the weights of elements was a tantalising indication that something was going on at a deeper, more fundamental level in atoms – that they possessed an internal structure. Everything made sense, claimed Prout in 1815, if nature's basic building block is the hydrogen atom and all heavier atoms are simply different numbers of hydrogen atoms glued together.

Another tantalising pattern in the properties of the elements was noticed in the late nineteenth century by a Siberian chemist called Dimitri Mendeleev. While preparing a textbook of chemistry, Mendeleev created a card for each of the 67 known elements, writing on each card the element's properties, such as its melting point and its chemical behaviour. To his surprise, he discovered that, if he placed the cards in horizontal rows, with most of the elements in order of increasing atomic weight, those with similar properties appeared magically in vertical columns. Mendeleev's periodic pattern in the properties of elements appeared to be telling scientists that atoms must be made of smaller things, just as surely as the pattern discovered by Prout in the weights of the elements.

At the close of the nineteenth century, a smaller building block of the atom did actually come to light. The Cambridge physicist J. J. Thomson used high voltage to rip it free of atoms. The 'electron', the long-sought carrier of electricity, was fantastically tiny. According to Thomson's measurements, it was a mere 2000th the mass of hydrogen, the lightest atom. It was too small to be Prout's subatomic building block. And neither was it clear whether it had anything to do with the pattern in atomic properties discovered by Mendeleev. But the electron did allow a first-ever stab at picturing what it was like inside an atom.

The electron carries a 'negative' electric charge. Nobody quite knows what electric charge is, only that it comes in two distinct types – negative and positive. Opposite charges attract while similar charges repel each other. Since in daily life we do not find the electric force tugging us this way and that, we know that, overall, matter must be electrically neutral, with its negative charge perfectly balanced by an equal amount of positive charge. In an atom, therefore, the negative charge of the electrons must be balanced by the positive charge of 'something else'. Despite not knowing what the something else was, Thomson still managed to concoct a plausible model of the atom, picturing it as a diffuse sphere of positive charge with tiny electrons studded in it like raisins in a Christmas, or 'plum', pudding.

Thomson's 'plum pudding' model was the accepted picture of an atom well into the twentieth century. But, in 1909, one of the giants of experimental physics changed everything. The New Zealand-born physicist Ernest Rutherford was a pioneer of the study of 'radioactivity'. The phenomenon had been stumbled on by the French chemist Henri Becquerel in 1896, when he had noticed that photographic plates were fogged by mysterious 'rays' coming from rocks containing uranium and thorium. Marie Curie had picked up the baton and, in Paris in 1898, recognised, correctly, that radioactivity was a property of atoms. The rays from a radioactive substance were so intense that they were detectable even if its atoms were present in tiny numbers. And Curie exploited this property to dramatic effect by discovering two elements hitherto unknown to science: polonium, which hit the world headlines in 2006 with the poisoning in London of Russian dissident Alexander Litvinenko, and radium.

The same year Curie made her discovery that radioactivity was a property of atoms, Rutherford, working in Montreal, found that radioactivity involved the emission by atoms of two distinct types of ray, which he christened 'alpha' and 'beta' rays. Beta rays were shown to be electrons, but it took until 1903 for Rutherford, at the time working in Manchester with a young German student called Hans Geiger, to collect enough alpha rays from a sample of radium to show that – bafflingly – they were made of helium atoms, the second lightest atom after hydrogen.[3] It seemed that, during the process of radioactivity, one kind of atom – radium – spat out another kind of atom – helium. It was yet another piece of evidence that atoms were made of smaller things.

Rutherford eventually solved the puzzle of radioactivity. A radioactive atom, he and chemist Frederick Soddy determined between 1901 and 1903, is simply a heavy atom that is unstable. It seethes with surplus energy. Eventually, it sheds the excess by violently expelling an alpha or beta particle and, in the process, disintegrating, or 'decaying', into an atom of a lighter element.[4] But Rutherford did not need to know what radioactivity was in order to come up with a way of 'seeing' inside the atom. In 1903, he had measured the speed of alpha particles expelled by radium and found it was an incredible 25,000 kilometres per second – fast enough to travel from one side of the Earth to the other in less than a second. A sample of radium was like a miniature machine gun, sputtering out subatomic bullets at ultra-high speed. It was perfect, Rutherford realised, for probing the interior of the atom.

Rutherford's idea was to fire a radium machine gun at a thin foil of a material. As they passed through the foil, some of the alpha particles would inevitably be deflected from their paths, enabling Rutherford to deduce the internal structure of the material's atoms from the manner in which they were deflected. It was just like bouncing tennis balls off a mystery piece of furniture and deducing, from the directions in which the balls ricocheted, whether it was a chair or a table or a Welsh dresser. In the quest to lay bare the interior of the atom, Rutherford's genius was to turn the atom on itself. He would use one type of atom – the helium atom spat out by radium – to image an entirely different type of atom.

An alpha particle was four times heavier than a hydrogen atom, which made it about 8,000 times heavier than an electron. Rutherford's expectation, therefore, was that the alpha-particle bullets from his radium machine gun would drill right through his thin foil. They were as likely to be deflected by the electrons inside an atom as a bullet by a cloud of gnats.

Rutherford left the experiment to Geiger and a student from New Zealand called Ernest Marsden. Their radium machine gun sputtered its stream of alpha-particle bullets at a thin foil of gold. Geiger and Marsden, who within a few years would be firing real bullets at each other across the Western Front, then measured the deflection of the alpha particles. As expected, there was essentially no deflection. Then, one day, Rutherford poked his head around the door of their laboratory and made a ridiculous suggestion. He told Geiger and Marsden to look for alpha particles coming backwards from their gold foil.

Seeing an alpha particle ricochet backwards would be like firing a bullet into a cloud of gnats and seeing it bounce back the way it had come. But the mark of geniuses – and Rutherford was the greatest experimental physicist of the twentieth century – is that they are always open to the unexpected, never allowing their theoretical prejudice to limit what nature might reveal to them. And Rutherford was rewarded. Three days after he made his suggestion, Geiger and Marsden burst into his office with the most incredible piece of news. Out of every 8,000 alpha particles fired at the gold foil, one bounced back. As Rutherford admitted: 'It was quite the most incredible event that has ever happened to me in my life.'

It took two years for Rutherford to make sense of Geiger and Marsden's astonishing result. For an alpha particle to run into something inside an atom and for that something not only to stop it dead in its tracks but send it flying back the way it had come, that something had to be far more massive than an alpha particle. It also had to occupy a tremendously small fraction of the volume of an atom. After all, it was so small a target that it got in the way of only about one in 8,000 alpha particles.

By 1911, Rutherford had used the observations to deduce the internal structure of the atom. Instead of tiny electrons embedded in a diffuse cloud of positive charge, as Thomson had imagined, electrons flitted about a tight knot of positive charge at the centre of the atom. Only if nature crammed a large positive charge into an extremely small volume could that charge exert a repulsive force so overwhelming that it made an alpha particle execute a U-turn. According to Rutherford's estimates, the tight knot of positive charge must account for a whopping 99.9 per cent of the mass of the atom. Rutherford's model of an atom was about as far from Thomson's plum-pudding model as it was possible to imagine. The atom was like a miniature Solar System, with electrons like planets orbiting about a Sun-like 'nucleus'.[5]

As Rutherford's Cambridge colleague, the novelist and physicist C. P. Snow, observed:

> As soon as Rutherford got on to radioactivity, he was set on his life's work. His ideas were simple, rugged, material; he kept them so. He thought of atoms as though they were tennis balls. He discovered particles smaller than atoms, and discovered how they moved or bounced. Sometimes the particles bounced the wrong way. Then he inspected the facts and made a new but always simple picture. In that way, he moved, as certainly as a sleepwalker, from unstable radioactive atoms to the discovery of the nucleus and the structure of the atom.

Rutherford had come to England by a fluke. His scholarship to Cambridge had actually been won by another New Zealander, ranked above Rutherford. At the last moment, however, the man had got married and the scholarship had been passed on to the next in line. Rutherford was a loud, boisterous, domineering man. However, later in his life, when he was Lord Rutherford, a Nobel Prize-winner and fêted as one of the greatest experimental physicists of all time, he could still be brought to tears by the thought of how his life, but for one chance event, could have turned out so very differently.

The huge shock to Rutherford and everyone else was the

size of the nucleus. It turned out to be about 100,000 times smaller than the atom itself. The atom, amazingly, was over-whelmingly made of empty space. In fact, atoms are so empty that if it were possible to squeeze out all the space in matter, the entire human race could fit in the volume of a sugar cube. But why do atoms contain so much empty space? Or, to put it another way, why are they so big compared with their ultra-tiny nuclei? These questions turn out to be inex-tricably bound up with another, far more fundamental ques-tion: Why do atoms exist at all? Because, according to the laws of physics, they should not.

The Laws of Electromagnetism

The relevant laws of physics are the laws of electromag-netism, as formulated by the Scottish physicist James Clerk Maxwell in the 1860s. As mentioned before, Maxwell man-aged to distil all electrical and magnetic phenomena into one compact set of equations.[6] His starting point was to imagine the force that a magnet exerts on a chunk of metal, for instance, as being due to a ghostly magnetic 'field' of force extending outwards from the magnet into the surrounding space. He similarly imagined an electric 'field' of force extending outwards from electric charges, such as those trav-elling along a wire as an electric current.

But Maxwell's theory did more than merely describe the behaviour of electric and magnetic fields. It contained a big surprise. Examining the equations he had written down, Maxwell noticed that they permitted the existence of a wave – an undulation travelling through the electric and magnetic fields. Such an 'electromagnetic wave' would spread throughout space much like a ripple on the surface of a pond. And it had a characteristic speed in empty space.

Maxwell noted with astonishment that it was the speed of light.

No one – apart from the electrical pioneer Michael Faraday – had had the slightest inkling that there might be a connection between electricity and magnetism, and light. They were such different phenomena. But there it was, plain to see in Maxwell's equations – a wave of electricity and magnetism which travelled through empty space at the speed of light. It did not need a genius to guess that the two things were one and the same. Light, Maxwell realised, must be an electromagnetic wave.

The pay-off would change the world beyond recognition. Maxwell's theory predicted that simply jiggling the electrical current in a wire would cause the current to broadcast electromagnetic waves, not light visible to the eye but a longer-wave form known as radio waves. Maxwell died in 1879, aged 48, but his prediction was confirmed by the German Heinrich Hertz in 1887. And, in 1901, Guglielmo Marconi succeeded in transmitting a radio signal across the Atlantic, ushering in the epoch of instant communication that has made the modern world possible.

Maxwell's theory says that electromagnetic waves are created by any electric charge that 'accelerates': that is, one that changes its speed or its direction or both. Jiggling an electric current in a wire accelerates the current-carriers – electrons – which is why such an 'aerial' broadcasts radio waves. But this poses a problem for the atom. Precisely the phenomenon that makes long-distance communication a possibility makes Rutherford's 'planetary' model of the atom an impossibility. The difficulty is that an electron circling the nucleus of an atom is continually changing its direction and so continually accelerating. And, as an accelerated charge, it ought to broadcast electromagnetic waves, just like a tiny radio transmitter. But the electromagnetic waves carry away energy from the electron. Sapped of energy in this way, it should spiral down to the nucleus within a mere hundred-millionth of a second. Atoms, in short, should not exist.

But they do.

Far from winking out of existence in a hundred-millionth of a second, atoms appear to have been stable for the billions of years the Universe has been around. They have persisted, in fact, for something like 1 followed by 40 zeros times longer than the laws of electromagnetism predict that they should. Until 1998 and the discovery of the cosmic 'dark energy', this was the biggest discrepancy between an observation and a prediction in the history of science.[7]

Rutherford was baffled. He had succeeded spectacularly in revealing the interior of the atom. However, in doing so, he had discovered a major conflict in physics. The experiment with the gold foil had shown that the atom was a tiny planetary system. Yet the theory of electromagnetism predicted that such a system was so unstable it could not survive for even the blink of an eye. It was a paradoxical situation, and it was nigh on impossible to see a way out. But one man did. He was a young Danish physicist.

Niels Bohr had come to England in 1911 after completing his doctorate in Copenhagen and had worked with J. J. Thomson and then Rutherford. He saw that Rutherford's planetary picture of the atom, supported by hard experimental evidence, was compelling. But he also saw that the laws of electromagnetism, which had made electric motors and dynamos possible, were equally compelling. His revolutionary resolution of the atomic paradox was as simple as it was daring. In 1913, he claimed that the laws of electromagnetism did not operate inside the atom. Electrons, as they orbited a nucleus, did not broadcast electromagnetic waves and hence they did not spiral into their nuclei. In short, the known laws of physics did not apply in the domain of the ultra-small.

Bohr's evidence for his revolutionary idea was simply that the known laws of physics said atoms could not exist, and yet they did. But he did not know what replaced the known laws of physics in the microscopic realm. He did not know why electrons did not spiral down into their nuclei. An explanation of that would come from a French physicist called Louis de Broglie.

Particles Behave as Waves

De Broglie knew of Einstein's proposal that light waves could behave as particles – photons – and of its confirmation by both the photoelectric effect and the Compton effect. But de Broglie went one, almost unbelievable, step further. In his doctoral thesis in 1923, he claimed that not only can light waves behave like localised particles but that particles such as electrons can behave as spread-out waves. All the microscopic building blocks of matter, according to de Broglie, had two faces. All shared a peculiar wave–particle duality.

De Broglie's idea of 'matter waves' was so fantastic that most physicists ignored it entirely. However, everything changed when Einstein read a copy of de Broglie's thesis. The father of the photon was bowled over by the idea and convinced that de Broglie was on to something. What was needed was a demonstration that a particle like an electron could behave like a wave. In practice, this meant demonstrating that electrons could interfere with each other, since interference was the defining characteristic of waves. It was a feat that was achieved in 1927 by Clinton Davisson and Lester Germer in the US and by George Thomson in Scotland. The irony was that Thomson was the son of J. J. Thomson. The father received the Nobel Prize for proving that the electron is a particle; the son would share the Nobel Prize for refuting this and showing that, actually, the electron is a wave.

It is because *all* microscopic particles behave like waves that the conclusion drawn from seeing your face in a window can be extrapolated to *all* particles. It is not just the photon but every denizen of the microscopic world that dances to the tune of randomness. Superpositions and every other weird quantum phenomenon are common to every last one of them.

De Broglie, in his thesis, did more than simply postulate that particles of matter act like waves, he spelled out how big those matter waves are. Their size is inversely proportional to their momentum, which is the product of a body's mass and its velocity. In general, big things moving around in the everyday world, like jumbo jets and even snails, have a great deal more momentum than tiny things moving about in the microscopic world of atoms. And since, according to de Broglie, the size of the wave associated with a body is inversely proportional to its momentum, it follows that the waves associated with everyday things are far tinier than those associated with particles such as the electron.

Take a baseball thrown by a pitcher at about 150 kilometres per hour. According to de Broglie's hypothesis, it behaves like a wave with a wavelength of only about 10^{-34} metres.[8] That's a trillion trillion times smaller than an atom. No surprise then that nobody noticed that matter had a wave-like character until well into the twentieth century. The wavelengths of big things in the everyday world are simply so mind-cringingly small as to be utterly undetectable. This is why we do not see people rippling down streets or constructively or destructively interfering with each other.

But consider an electron travelling at about 6,000 kilometres per second. Being so light, it can easily be boosted to such a speed by a modest voltage of just 100 volts. Such an electron has a wavelength of about 10^{-10} metres. The significance of this is that it is comparable to the spacing between atoms in a material like a metal. So, if such electrons are fired at a metal, there is a good chance of seeing wave effects such as interference. This was exactly the strategy used by Davisson and Germer, and Thomson, to demonstrate the wave nature of electrons. They fired a beam of fast electrons at a metal target. The atoms in the metal are arranged regularly, evenly spaced in parallel layers like a stack of pancakes. So, when electrons are fired at the metal, some bounce off the surface layer. Others penetrate to the second layer before bouncing back out of the metal. Others to the third layer. And so on. But the key thing is that all the electrons reflected by the metal behave like waves. Consequently, there will be directions in which the electron waves from all the different layers are in step and so constructively interfere. And there will be directions in which they are completely out of step and so destructively interfere. It is necessary only to detect the numbers of electrons coming back from the metal in different directions.

This is exactly what Davisson and Germer did in the US, and Thomson in Scotland. And what they found was that in some directions there were lots of electrons flying off the metal, while in others there were absolutely none. And the directions in which there were lots of electrons alternated with the directions in which there were none. It was an interference pattern – or, strictly speaking, a 'diffraction pattern', a closely related phenomenon – incontrovertible proof that electrons do indeed behave like waves. It must have been an amazing thing to see. After all, it is one thing for an ivory-tower theorist like de Broglie to postulate the existence of something as ridiculously mad as matter waves, but it is quite another actually to 'see' electrons, which had been considered to be tiny billiard balls, behaving like ripples on a pond.

Waves Need Elbow Room

It is not immediately obvious how de Broglie's idea of matter waves explains why an electron in an atom does not spiral down to nuclear oblivion. However, a wave, being a fundamentally spread-out thing, needs elbow room.[9] The electron, as the lightest known subatomic particle, tends to have the biggest associated wave. This means that it is the particle most dominated by weird quantum-wave effects. And it also means that it needs more elbow room than any other particle. At the speed an electron typically travels inside an atom, in fact, its associated wave is as big as an atom. It defines the size of an atom.

There is one slight subtlety. Being about 2,000 times bigger than an electron, the nucleus of a hydrogen atom might be expected to have a wave a 2,000th the size of an electron wave. But in fact, the nucleus is more like 100,000 times smaller. The discrepancy arises because the electron is constrained by the electric force, whereas particles in the nucleus are governed by a far stronger nuclear force. The stronger the force, the faster a particle moves, which means the momentum of the nucleus is higher than might be expected and its wavelength that much smaller than a 2,000th the wavelength of an orbiting electron.

So the reason electrons do not spiral down into their nuclei is because electrons have comparatively large waves associated with them and such waves need elbow room. It is the reason atoms exist. But what exactly prevents an electron wave being squeezed down into a small space? In short, what pushes back if electrons are squeezed too close to their nuclei? What is responsible for the resilience, the stiffness of matter? For an answer, it is necessary to switch back to thinking of electrons as particles and reconsider the double-slit experiment again.

The Heisenberg Uncertainty Principle

Recall that when photons are fired at two closely spaced, parallel slits in an opaque screen, a pattern of vertical stripes appears on a second screen placed at some distance beyond the first. The stripes consist of bands struck by lots of photons alternating with bands studiously avoided by them. Such an 'interference' pattern makes sense if, associated with the photons, are quantum waves which inform the photons where they should end up. The waves emerging from one slit overlap with the waves emerging from the other, periodically reinforcing and cancelling each other out to create the distinctive zebra stripes of photons on the second screen.

Of course, in the light of de Broglie's insight, it is clear that it is not just particles of light that can be made to interfere with each other by firing them at slits in an opaque screen. The double-slit experiment will work with electrons or atoms or any other particles, although, in practice, the more massive the particles, the shorter their wavelength and the more difficult it is to get them to interfere. Or, if you can arrange for them to do so, the more difficult it is to see the zebra-striped pattern.

Whatever the particle, recall that interference requires two things to mingle. So, when particles are fired at the two slits one at a time, with long intervals in between, the slow build-up of an interference pattern on the second screen implies that each particle goes through both slits simultaneously – that it is in two places at once.[10] But what if we discover precisely which slit each particle goes through? Clearly, if we can do this, the interference pattern will vanish, since we will have ruled out the possibility of each particle going through both slits simultaneously and mingling with itself.

Now, the vanishing of the interference pattern has profound and disturbing implications for particles of matter – not to mention the nature of ultimate reality. And this can be seen by imagining precisely how we might discover which slit a particle goes through. Imagine inflating the experiment in size so that instead of dealing with photons or electrons or other subatomic particles, we are dealing with machine-gun bullets, and the screen is a thick sheet of steel – a couple of centimetres thick, say – so that the vertical double slits are like two deep channels gouged in the steel sheet. Now focus on those bullets. As they go down the channels they ricochet off the walls, and as they do, the walls – and the whole sheet of steel – recoil. This provides a means of detecting which slit a bullet goes through.

Imagine, for simplicity, bullets going through the slits, bouncing off the walls and ending up peppering the mid-point of the interference pattern. In this case, we can say that if the steel sheet recoils to the left, a bullet must have gone through the left slit. And if the sheet recoils to the right, the bullet must have gone through the right slit. So we now know that if we do not detect which slit each bullet goes through, we get the zebra stripes on a second screen – bands which are peppered with bullets alternating with bands which are hit by no bullets. And if we do – noting the recoil of the steel sheet – the zebra stripes must vanish.

Focus on the zebra stripes. What must happen for them to be washed out? Well, all that is required is for a bullet that was destined for a bullet-riddled band to hit either the bullet-riddled band or the bullet-free band adjacent to it. This will be enough to make bullets pepper the second screen evenly, smearing out the zebra stripes into a uniform grey. What we are talking about is each bullet as it flies through the air having a bit of random sideways jitter, so there is enough uncertainty in where it ends up to wipe out the interference pattern. This sideways jitter can only have been imparted to the bullet when it ricocheted off the wall of the channel gouged in the steel sheet.

So what is happening here is that the act of trying to locate which slit a bullet goes through endows it with exactly the sideways jitter necessary to destroy the interference pattern. This jitter is nature's way of protecting quantum theory. To behave in a wave-like way, a particle must be able to do two things at once, so that the waves associated with those indistinguishable possibilities mingle or interfere. If you can distinguish between those possibilities – by determining that one thing has happened rather than the other – then there are no longer two indistinguishable possibilities to interfere. And the thing that your measurement does that destroys the possibility of interference between particles is add random jitter to those particles.[11]

To put it more precisely in the machine-gun example, locating the slit a bullet went through – which is synonymous with locating exactly where the bullet is – adds a random jitter, or uncertainty, to its velocity, or momentum. And this is the crux of the matter. As the young German physicist Werner Heisenberg discovered in 1927, there is a trade-off between two things: the more sure we are of the location of a particle, the less well we can know its momentum. And the reverse is true too: the more sure we are of the momentum of a particle, the less well we can know its position.

And this is fundamental. We are talking about irreducible uncertainty in our knowledge of subatomic particles, just like the irreducible unpredictability in their behaviour. In the everyday world we can know that a person is at a cross-roads in the city and walking at three kilometres an hour. In the microscopic world we can never know both these things with certainty. Knowing one very precisely unavoidably means that we must be ignorant of the other. There is an ultimate limit on our knowledge of the world. Look too closely at reality and there is nothing sharply defined to see. It dissolves into a meaningless blur, much like a newspaper photograph dissolves if examined too closely.

And it is this 'Heisenberg uncertainty principle' which ultimately explains why atoms do not shrink down to nothing and why the ground beneath your feet is solid. Granted, the fact that electrons are waves and waves need elbow room gives a partial explanation. But the rest of the explanation comes from considering what happens if an electron is forced close to a nucleus. This means that its location becomes very precisely known. But, according to the Heisenberg uncertainty principle, the more sure we are about a particle's location, the more unsure we must be about its momentum. It is pretty much like having a bee in a box. Shrink the box and the bee gets angry and batters ever more violently on the walls of its prison. As it is for bees in boxes so it is for electrons in atoms. Atoms, in the words of the poet Adrian Mitchell, 'rock and roll all over the place'. When you tread on the ground, your weight compresses the atoms of which it is composed. This squeezes the electrons in those atoms marginally closer to their nuclei. And the Heisenberg uncertainty principle causes them to resist, to push back. This is why the ground is solid, why matter is stiff. Yes, it is because of the wave nature of electrons. But it is also because of the irreducible uncertainty of the microscopic world, the fact that there is a limit to our knowledge of ultimate reality. This is ultimately what the solidity of the ground under your feet is telling you.

3

No More than Two Peas in a Pod at a Time

How the variety of the world is telling us that there must be something that prevents electrons from piling on top of each other

'It is the fact that electrons cannot get on top of each other that makes tables and everything else solid.'

Richard Feynman

'Quantum mechanics? What do they do all day?'
Rob Evans (performance poet)

Look about you at everything from a dandelion to a hurricane brewing in the Gulf of Mexico to a newborn baby to a star twinkling in the twilight sky. One of the most striking features of the world around us is its extraordinary, boundless variety. As guessed so presciently by Democritus two and a half millennia ago, all this incredible variety is merely a reflection of the vast number of ways it is possible to link together a small number of building blocks, or atoms. Out of simplicity, paradoxically, comes complexity. Everything is in the combinations.

The variety of the world is therefore telling us that there cannot be just one kind of atomic building block – there must be many. But why are there many rather than one? The reason must have something to do with what makes one kind of atom different from another. And what makes atoms different is the number of electrons they contain. It is those electrons, orbiting far from the central nucleus, that are the atom's interface with other atoms. They define its 'surface' and how it links up with other atomic Lego bricks. In short, they are what makes a calcium atom calcium, a gold atom gold and a platinum atom platinum.

So the boundless variety of the everyday world is telling us something crucial about electrons. In fact, that crucial thing is that electrons have a remarkable and powerful aversion to each other. But that is jumping the gun a bit.

To appreciate why the variety of the everyday world is telling us such a peculiar and specific thing requires some background. In particular, it is necessary to know something about the way electrons arrange themselves inside atoms and how this creates atoms which behave differently from each other.

Like all particles of matter, electrons behave like waves. According to de Broglie, the smaller the momentum of a particle, the bigger the wave. Being nature's lightest particle with mass,[1] the electron therefore in general has the biggest wavelength. This is of course why, of all the subatomic particles, it displays the most striking wave-like behaviour, and why it is totally impossible to understand the atom without taking this aspect of its nature into consideration. Recall that it is an electron wave's desire for elbow room that prevents it from spiralling down to nuclear oblivion and so makes atoms possible at all.

The probability wave associated with an electron in an atom is a bit like a sound wave in an organ pipe. Such a wave, confined by the walls of the pipe, can vibrate in only a limited number of ways, each of which is characterised by a distinct pitch, or 'frequency'. Similarly, an electron wave in an atom, gripped by the electrical force of the nucleus, can vibrate at only a limited number of frequencies.

An organ pipe has a lowest, or fundamental, frequency plus higher-frequency 'overtones'. The higher the frequency, the more peaks and troughs of the wave in a given space – the choppier and more violent it is. For an electron in an atom, such a wave corresponds to a faster-moving, more energetic particle, one able to defy the electrical attraction of its nucleus and orbit at a great distance.

The fact that an electron wave can vibrate at only a limited number of frequencies means that an electron in an atom is not free to orbit at an arbitrary distance from the nucleus. Only certain special distances are permitted – and none in between. Say the laws of physics allowed you to stand only three metres from a tree or eight metres or 27 metres but not any other distance. Ridiculous as it seems, this is pretty much the way it is for electrons orbiting an atomic nucleus.

The innermost orbit permitted for an electron in an atom is just the one fixed by the Heisenberg uncertainty principle – by the electron's hornet-like resistance to being confined in too small a space.[2] This corresponds to the lowest possible vibration of the electron wave – the fundamental frequency. The other permitted orbits, at greater and greater distances from the nucleus, correspond to the higher-frequency over-tones.

Not surprisingly, the innermost orbit is labelled with a 1, while successively more distant orbits from the nucleus are referred to as 2, 3, 4, and so on. These 'quantum numbers' are yet one more example of the way in which just about every-thing in the microscopic realm of atoms is grainy rather than continuous, meted out in discrete chunks, ubiquitous 'quanta'.

Actually, there is another subtlety in the way electrons orbit in atoms. The electron probability wave can be a quite complicated three-dimensional thing. Consequently, it may correspond to an electron that is not only most likely to be found at a particular distance from the nucleus but also more likely to be found in particular directions rather than others. For instance, an electron wave might be bigger over the north and south poles of an atom than anywhere else, making these the most likely places to find the electron.

Once again, it is evident that the words used to describe the everyday world of the big are simply not applicable in the realm of the very small. Although Rutherford painted a vivid picture of electrons orbiting an atomic nucleus like planets around the Sun, electrons actually orbit in a very un-planet-like way. Not only can they orbit only at certain special distances from the nucleus – and such locations are no more than the 'most probable' places to find them – but they also may have a tendency to be found in certain directions rather than others. In recognition of this, scientists tend not to talk of electron orbits at all. Instead, they use the term 'electron orbital' to describe nature's weirder, more complex reality.

Describing a direction in three-dimensional space needs two numbers; think of 'latitude' and 'longitude' on a globe.[3] Consequently, an electron wave whose height changes with direction in space requires two more quantum numbers to pin it down, as well as the one specifying its distance from the nucleus. A total of three.

Now nature permits no more than two electrons to have an electron wave described by a particular trio of quantum numbers. And this peculiarity, it turns out, is the absolute key to the endless variety in the world around us. But to appreciate why still requires some more background on the way in which electrons are arranged inside atoms.

All the orbitals at a particular distance from the nucleus – that is, with the same principal quantum number but different directional quantum numbers – are said to compose a 'shell'. The maximum number of electrons that can occupy the innermost shell, labelled with a 1, turns out to be two. The maximum number that can occupy the next shell further out, labelled 2, is eight. For the one after that, labelled 3, the number is 18. And so on.

Here – finally – we are getting to the nitty-gritty of what makes one kind of atom different from another. Recall that different kinds of atoms have different numbers of electrons. The lightest atom, hydrogen, has one, and the heaviest naturally occurring atom, uranium, has 92. Imagine what happens if, hypothetically, electrons are added, one at a time, to make atoms of successively heavier elements. The first available shell is the innermost one, nearest to the nucleus. As electrons are added, they first go into this shell. When it is full and can take no more electrons, they pile into the next available shell, further away from the nucleus. Once that is full, they start filling up the next most distant. And so on.

A hydrogen atom has one electron in the innermost shell and an atom of the next heaviest element, helium, has two. This is enough to fill the shell to its capacity. The next biggest atom is lithium, with three electrons. Since there is no more room in the innermost shell, the third electron starts a new shell further out from the nucleus. The capacity of this shell is eight. For atoms with more than ten electrons, however, even this shell is all used up, and another still further from the nucleus begins to be filled.

Recall that Mendeleev found that if he placed cards with the names of the elements in horizontal rows, with most of the elements in order of increasing atomic weight, those with similar properties appeared, magically, in vertical columns. This 'periodicity' in the properties of atoms turns out to reflect the periodic filling up of shells by electrons in atoms. In particular, it reflects the number of electrons which are left in the outer shell of an atom. All atoms with one electron in their outermost shell, such as lithium, sodium and potassium, have very similar properties. So too do all atoms with two electrons in their outermost shell, such as magnesium, calcium and radium.

The reason for this is that it is the electrons orbiting furthest from the nucleus that are the ones that come into contact with other atoms. If you think of an atom as a billiard ball, it is the outermost electrons that define its 'surface', that give it its characteristic size. And being on the outer rim of the atom, they also determine how it joins up with other atoms. Imagine the outer electrons as hooks with which an atom can latch onto other atoms. The picture is crude, but an atom with one electron in its outermost orbit, such as sodium, as found in table salt, hooks up with other atoms in one particular way. An atom with two electrons in its outermost orbit, such as calcium, found in your bones, latches on in another way. An atom with three outer electrons, such as aluminium, the lightweight metal, in yet another way. And so on.

The directions in space in which these outermost electrons are most likely to be found determine precisely how one type of atom sticks together with other types to make compounds such as polythene or ammonia or methane. Chemists picture the preferred directions of electron waves as spine-like 'bonds' extending hedgehog-like from an atom and capable of linking up with the hedgehog-like bonds of another atom. Chemistry, it turns out, is ultimately electron geometry.

The most stable atoms turn out to be those whose outer shells are completely filled with electrons. Since they have no electronic spines sticking out, they have no desire to marry themselves with other atoms. They are content with their lot. They are aloof. They are complete. And it is the desire of atoms to achieve this exalted complete state that determines almost all of chemistry. For instance, an atom of chlorine, which is one electron short of filling up its outer shell, will grab an electron from an atom of sodium, which has just one electron in its outer shell. After this game of give and take, both atoms will end up with filled outer shells. The compound created by this marriage of convenience is none other than sodium chloride, or common-or-garden table salt.

But there are also other ways to achieve electronic nirvana. Instead of one atom borrowing an electron and another atom donating it, two atoms can share their outer electrons so that each has the illusion of its own filled outer shell. The most important example of this for carbon-based creatures like us is, well, carbon. Since an atom of carbon has four electrons in an outer shell capable of a maximum occupancy of eight, there is a big incentive to join up with other carbon atoms. Four plus four equals a full house of eight. It is this tendency of carbon atoms to enter into same-sex relationships – in fact, multiple same-sex relationships – that is behind the existence of a bewildering range of long-chain carbon 'molecules', most importantly the molecules of life such as the sweeping double helix of DNA.

Apologies for the gory detail on how electrons arrange themselves in atoms, but there is no other way. The variety of the world stems from there being not one kind of atom but many. And the fact there are many types of atoms stems from the fact that atoms have a very particular internal structure – that, within an atom, there exist concentric shells, each capable of containing a fixed number of electrons, with the electrons in an incomplete outer shell determining how the atom behaves, whether it is calcium or uranium or gold. And, ultimately, the reason atoms have this particular structure – as mentioned before – is because of the unsociability of electrons.

Imagine the electron orbitals as the rungs of ladder, with the lowest-energy, innermost orbital corresponding to the lowest rung. Adding electrons to make heavier and heavier atoms is therefore like adding electrons to the bottom rung of the ladder, then, when this is filled up, to the second rung, and so on. Now there is a tendency for things to head for the lowest-energy state, just as surely as a ball rolling to the foot of a valley, the state in which the ball has the lowest 'gravitational energy'. But in an atom this would mean the electrons – whether there was one or 92 – all heading for the bottom rung of the ladder, the lowest-energy orbital.

If this happened in atoms – if all the electrons crowded together in the lowest orbital – there would be no such thing as an electron shell with an occupancy limit never to be exceeded. And if there were no electron shells, the idea of a filled-up shell would clearly be meaningless. With no desire on the part of atoms to acquire a filled-up outer shell, the driving force behind all atomic link-ups would be gone. All types of atom would behave in the same antisocial way. There would be no variety. No difference. No us.

So you see, what the variety of the world is actually telling us is that there must be something that prevents electrons piling on top of each other, a hitherto unsuspected law of nature – one that somehow explains the internal structure of atoms. There is. It is called the Pauli exclusion principle, after the Austrian physicist Wolfgang Pauli, who proposed it in December 1924.

The Pauli Exclusion Principle

Pauli, aged 21, exploded onto the scientific scene in 1921 when he published a book on relativity which taught even Einstein a thing or two about his own theory. Famously forthright – or just plain arrogant (physicists were divided on this) – Pauli was not averse to standing up in a lecture and telling the lecturer they were talking total garbage, irrespective of who they were and their reputation. So legendary was Pauli's cockiness that there was a joke in the physics community that went something like this: Pauli dies and goes to heaven. God asks him if there is anything about physics he would like to know. Pauli replies he would like to know why the fine-structure constant, which governs the strength of the interaction between light and matter, has the value $1/(137.035999071 \ldots)$ rather than simply $1/137$. God goes to the blackboard and begins scribbling equations. After a short while, Pauli's face breaks into a triumphant grin. He grabs the chalk from God and says: 'That's it – that's where you went wrong.'

Despite his dodgy ego, Pauli was one of the towering figures of twentieth-century physics. In 1930, he famously predicted the existence of the 'neutrino', a ghostly particle that carries away the missing energy in radioactive beta decay. So slippery is the neutrino that 100 million million solar neutrinos pass through you every second, utterly unhindered by the atoms of your body. The neutrino would have been enough to make his name. But it was the exclusion principle, for which he won the 1945 Nobel Prize for Physics, for which Pauli was most famous.

The Pauli exclusion principle is one of the most remarkable edicts in the Universe, yet it is notorious for defying the best attempts of physicists to explain it in everyday language. But never fear. The first step in understanding it is to appreciate something about the double-slit experiment: namely, that one particular conclusion drawn from it is actually more general than it at first appears.

Recall that if the slit through which each particle goes is identified, no zebra-striped interference pattern builds up on the second screen. Instead, the particles that go through the slits pepper the second screen uniformly. An examination of how locating the slit a particle goes through smears out the interference pattern leads to the conclusion that the act of measurement adds random sideways jitter to the particle as it flies through space. This jitter, like so many things in the quantum world, is fundamental, intrinsic, irreducible. It tells us there is an unavoidable constraint on how well we can know both the location and momentum of a microscopic particle. The more precisely we pin down the location, the more uncertain is our knowledge of its momentum. And vice versa. It is a trade-off.

From the wave rather than the particle point of view, this 'uncertainty principle' is actually trivial. The more localised a wave is in space, the more choppy and violent it is, and consequently the more energy and momentum it carries.

The uncertainty principle acts to protect interference, the basis of so much quantum weirdness. If two options are open to a microscopic particle and it is possible to discover – even in principle – that it has taken one rather than the other, then this rules out the possibility of interference, since a fundamental prerequisite of interference is that two things must mingle. If, however, it is not possible to identify which option has been chosen, interference between the waves representing the two options will occur.

This is the key point – the generalisation of the double-slit result. Interference occurs if two options are indistinguishable.

What has this got to do with electrons? Well, electrons, it turns out, are fundamentally indistinguishable. Once again, we are talking about a property of the microscopic realm with absolutely no parallel in the everyday world. We may say that two Barbie dolls are indistinguishable but actually, at a detailed molecular level, they are not. In fact, even at a grosser level, one doll may have a few more hairs on its head than the other or a few more crumples in its outfit. In the everyday world, no two objects are truly identical. Contrast this with the microscopic world. As far as we know, every one of the trillions upon trillions of electrons in the Universe is absolutely identical. It has no scratches on it, no scars, no blemishes, nothing whatsoever to distinguish it from any other member of the tribe of electrons. And this indistinguishability is truly something new under the sun.

The key thing, remember, is that things that are indistinguishable can interfere with each other. And since it is impossible to distinguish one electron from another, this has consequences for atoms, since they contain electrons.

Imagine there is a process that involves two identical particles which interact with each other. The particles could be any two that are indistinguishable. For instance, they could be two electrons or two photons or even two gold atoms (it is best to keep things as general as possible at this point rather than plump for the example of two electrons). And, in the most general case, the detailed nature of the interaction between the particles is unknown. They might come together, collide and simply bounce off each other. Or they might do any of a number of other things. The point is, we are ignorant of the details.

Let us assume that, as in the double-slit experiment, we have access to the particles only before and after they interact. Now, say the two particles start off at location 1 and location 2, respectively. They then interact and end up at locations 3 and 4. There are two possible ways this could happen. The particle which started out at location 1 could end up at location 3, and the particle which began at location 2 could end up at location 4. Or the particle which started at location 1 could end up at location 4, and the one which began at location 2 could end up at location 3.

Of course, we could tell which of these possibilities actually happened if the two particles were different – if one were green and one were blue, or if one were tattooed 'Particle A' and the other 'Particle B'. But the two particles are absolutely, utterly indistinguishable. In practice, therefore, it is impossible to tell which of the possibilities actually occurred. This is the new thing that indistinguishable particles bring to the party. Their indistinguishability means that the events they participate in may also be indistinguishable. And, in the microscopic world, this has an important consequence because, as already emphasised, if two events are indistinguishable, the probability waves representing each possibility can interfere with each other.[4]

In the case of the two identical particles starting off at locations 1 and 2 and ending up at locations 3 and 4, it is possible to be more specific. The total wave height for the process – which, recall, has to be squared to give the process's probability – equals the wave height for the first possibility plus the wave height for the second possibility. Now if the chance of one person throwing a six on a dice is ⅙ and the chance of someone else tossing a coin and it coming up heads is ½, then the chance of both these things happening together is ⅙ × ½ = ¹⁄₁₂. And this is precisely how the heights of the waves behave when dealing with identical particles. The total height for one particle going from location 1 to 3 and the other from location 2 to 4 is therefore $H(1 \text{ to } 3) \times H(2 \text{ to } 4)$. So the wave height for the whole process, which involves both possibilities, is $H(1 \text{ to } 3) \times H(2 \text{ to } 4) + H(2 \text{ to } 3) \times H(1 \text{ to } 4)$.

Now there is something that needs to be pointed out about the height of the quantum wave associated with an event. Like all waves, it requires two numbers to describe it. One is needed to denote the maximum height, or 'amplitude', of the wave. However, since the wave undulates up and down and does not always have this height, another number, known as the 'phase', is needed to pin down the location of the peaks.

An easy way to visualise the height of the quantum wave is as an arrow pointing in a particular direction in space, just like the hand of a clock. The arrow has an 'amplitude' – simply the length of the hand. And it also has a 'phase'. This is defined with respect to some reference direction, such as 12 o'clock. In this picture, the height of the wave is simply the height of the tip of the hand above the zero level – in the case of the clock defined by the line joining 9 o'clock and 3 o'clock.

Back to the two indistinguishable possibilities involving those two indistinguishable particles. Say locations 3 and 4 are at the same place. The wave height for the whole process is therefore $H(1 \text{ to } 3) \times H(2 \text{ to } 3) + H(2 \text{ to } 3) \times H(1 \text{ to } 3)$. In other words, the height of the quantum wave for the event is the sum of the height of the quantum wave for the possibility when the particles are one way round and when they have exchanged places.

Say the end location is the same distance from location 1 as it is from location 2. This makes the two indistinguishable possibilities mirror images of each other. And if this is the case, it is reasonable to suppose that the probability of each possibility must be the same – which is to say the square of the height of the wave representing each possibility must be the same.

Now arrows with the same length have the same square, irrespective of the direction in which they are pointing. Think of the hand of a clock. The square of its length is the same whether it is registering 2 o'clock or 9 o'clock. So you can imagine the arrows representing the quantum wave of each of the two possibilities as two hands of equal length on a clock face.

Now here is the point. It does not matter what the angle between the arrows is – the squares will always be the same. Say arrow two, representing possibility two, is x degrees ahead of arrow one. Imagine swapping the ingoing particles at locations 1 and 2. Arrow one now looks like arrow two. In other words, it has been rotated by x degrees from its original direction. Now swap the two outgoing particles. The same thing happens. Arrow one is rotated another x degrees from where it was – a total of $2x$ degrees. But exchanging the ingoing and outgoing particles simply gets things back to where they started. It restores the initial situation. So $2x$ degrees must equal a complete turn, since rotating something through a complete turn leaves it looking the same. Or

69

two turns. Or three. And so on. They all leave an arrow looking the same.

Consider the possibilities. If $2x$ equals a complete turn, then x must equal half a turn. If $2x$ equals two full turns, then $x =$ one turn. If $2x$ equals three full turns, then $x =$ one and a half turns. If $2x =$ four full turns, then $x =$ two turns. If $2x =$ five turns, then $x =$ two and a half turns. And so on. But rotating something by one and a half or two and a half turns is the same as rotating something by half a turn. And rotating something by two or four turns is the same as rotating it by one turn. So it is clear there are actually only two possibilities. The probabilities of the two events are unaffected if the arrows representing the heights of the waves for each are either half a turn or one turn apart.

What does this mean in the real world? Take the second possibility first. If the arrows are one turn apart, then clearly they point in the same direction, and so add up. Think of travelling five kilometres to the north-west along an arrow and then five kilometres to the north-west along a similar arrow. That would be the same as travelling to the north-west along an arrow ten kilometres long. So if the arrows are one turn apart, the height of the wave is doubled, which means the probability of the event happening is four times bigger than either event happening on its own.

What this means is, whatever the probability of one particle ending up at a particular end location, the probability of two ending up there is four times bigger. Naively, you might expect that the probability would be only twice as big. With identical particles, it appears the probability is enhanced. The fact one particle is in a particular location enhances the probability of another identical one also being found there. And the result is actually even more general than this. The fact that one particle is in a particular 'quantum state' – that is, doing a particular thing – enhances the probability that another will do the same thing. It is follow-my-leader, sheep-like behaviour. One sheep heads for a tree at the end of a field. Then another joins it. And another. In no time, the whole flock is heading for the same tree.

This sheep-like behaviour is behind the operation of a laser. Once an atom emits a photon of a particular frequency, moving in a particular direction, there is an enhanced probability that another atom will emit a photon of the same frequency, travelling in lockstep. And when there are two photons, the probability of a third joining is enhanced. In no time at all, the result is an avalanche of photons, all flying through space with identical properties. This 'stimulated emission' produces light waves that are in lockstep, their crests and troughs perfectly lined up, and this is the reason for the laser's unprecedented brightness.

So much for one of the possibilities open to two interacting indistinguishable particles. What about the other possibility, in which the arrows are half a turn apart? Well, if the arrows are half a turn apart, they point in opposite directions, and so cancel each other out. Think of travelling five kilometres to the north-west along an arrow and then five kilometres to the south-east along an arrow pointing in the opposite direction. You would be back where you started. So if two arrows are half a turn apart and cancel out, the height of the wave is zero. The event has no probability of occurring. It does not happen. Period.

If two identical particles behaved like this, there would be no way they could end up in the same location. More generally, they could not even do the same thing. Far from displaying gregarious, sheep-like behaviour, they would be antisocial, with an overwhelming aversion for each other. This aversion is known as the Pauli exclusion principle.

So, remarkably, the mere fact that two particles are indistinguishable leads – because of the interference of indistinguishable possibilities – to two strikingly different behaviours. On the one hand, identical particles can be antisocial; on the other, they can be gregarious. The question is: does nature avail itself of the two possibilities open to it? Are there particles that display gregarious, sheep-like behaviour and particles that are deeply antisocial? The answer is yes. Nature's fundamental particles really do fall into two distinct camps. Those that are gregarious are known as 'bosons', and those that are antisocial are known as 'fermions'. So what determines whether a particular particle is a boson or a fermion? The answer is its 'spin'.

Spin and Why It Matters

Spin is yet another of those quantum properties with no analogue in the everyday world. Despite the image it conjures up of an ice-skater whirling on the spot, spin actually tells us what a particle looks like when viewed from different directions – or, equivalently, what it looks like if you rotate it. Like everything else in the microscopic world, from electric charge to visible light, it comes in discrete chunks. In other words, there is a quantum of spin. A particle with twice the basic unit of spin, or spin 2, looks the same if you rotate it through half a turn – think of a double-headed arrow. A particle with spin 1 looks the same if you rotate it through one turn – think simply of a normal arrow. And nature does not stop there. It permits the existence of a particle with spin ½ (for technical reasons, the quantum of spin is actually half the standard unit of spin). Such a particle – and this is scarcely believable – looks the same only after it has been rotated through two turns.

If quantum spin is something new under the sun, spin ½ is something *doubly new* under the sun. Imagine not being the same person if you turn round once but only if you turn round *twice*. Well, that is the way it is for electrons, the most common example of a particle with spin ½. Specifically, after one turn, the arrow representing the height of the quantum wave of an electron points opposite to its initial direction. Only after two turns does it come back to pointing the way it was at the start.

But what has spin got to do with particles being antisocial or gregarious, with obeying the Pauli exclusion principle or not? This is the key question.[5]

Think of those two identical particles coming together and interacting at the same spot. Recall that because the particles are indistinguishable, the height of the quantum wave for the event is the sum of the height of the quantum wave for the possibility when the particles are one way round and when they have exchanged places. The two possibilities can be represented by two arrow-like hands on a clock face. And nature permits two situations: the arrows can point in the same direction and add up, or they can point in opposite directions and cancel. The latter leads to the Pauli exclusion principle – zero probability for two particles being in the same place or doing the same thing.

So what happens when two electrons, with spin ½, exchange positions? Think of two electrons side by side as two identical footballs. Since it is important to keep track of their orientations, imagine the two footballs are lying side by side in an east–west direction, with little red flags sticking out to the west. Now make the two footballs change places.

And do it in the following, rather odd, way. First, roll the westerly football round the surface of the easterly one (assume the flag can survive being squashed). This causes its red flag to go from pointing west to north to finally pointing to the east. In other words, the easterly football goes through a half turn clockwise. Now imagine the two footballs back in their original positions and a similar manoeuvre being performed on the easterly football. Roll it round the westerly one. This causes the red flag to go from pointing west to south to finally east. In other words, the easterly ball goes through a half turn anticlockwise.

The net effect of interchanging the two footballs, therefore, has been to rotate one through a complete turn with respect to the other. But remember that for a spin-½ particle, it takes two complete turns to get the arrow representing the height of its wave to where it was at the start. After only one turn it points in the opposite direction. But this is precisely what is required to cancel out the two possibilities in the case where two identical particles come together and interact. Spin-½ particles like the electron must therefore be nature's fermions. They must be the antisocial particles – the ones which obey the Pauli exclusion principle. And spin-1 particles, whose arrows come back to their start position after one turn, so the possibilities do not cancel out when two identical particles come together, well, they must be nature's bosons – the gregarious particles.[6]

Atoms are the way they are because electrons are fermions which obey the Pauli exclusion principle.[7] Try and put two electrons together, and they resist with all their might. It is this tremendous aversion to each other, this desire to avoid each other at all costs, which stops them all piling on top of each other. The Pauli exclusion principle prevents more than one electron from occupying the same quantum state. So the inner shell of an atom can contain only one electron, the next shell four, the next nine, and so on. But wait a minute. Isn't the maximum occupancy of the inner shell two, of the next eight, and the one after that 18? Right. It turns out that the Pauli exclusion principle prevents identical particles being in the same location. However, electrons have a way of being non-identical. It's all down to their spin again.

An electron with spin, in common with all moving electric charges, acts like a magnet (despite the fact that its spin is intrinsic and it is not actually spinning). In fact, spin is responsible for the magnetism in iron and for boosting the magnetic field inside an electrical coil, which makes possible the electric motors in hairdryers and food mixers, and the dynamos that generate the world's electrical power. The manipulation of the spin of electrons by magnetic fields has also made it possible to store and retrieve vast amounts of data on the disk drives of computers and iPods.

In a magnetic field, the spin of an electron behaves like a tiny compass needle. Actually, it behaves like a *quantum* compass needle. Unlike a familiar compass needle, it cannot align itself in any direction whatsoever but only in two possible directions: along the direction of the field or against it.[8] Think of the two possibilities as an electron spinning clockwise or anticlockwise. Well, a clockwise and an anticlockwise electron are not identical and so can share the same location in space, or quantum state. This is why each shell in an atom can contain twice as many electrons as might be expected.

Now it is possible to elaborate on the explanation given before for why the ground beneath your feet is solid,[9] which was that your weight squeezes the atoms but the electrons in those atoms resist being squeezed into a smaller space by buzzing about faster like angry bees. But while this effect, due to the Heisenberg uncertainty principle, explains the existence of all atoms and is the entire explanation for the resistance to squeezing by the simplest atom, hydrogen, with its solitary electron, for all heavier atoms another factor comes into play. And that is the Pauli exclusion principle. No more than two electrons can share the same quantum state. No more than two peas in a pod at a time. When your weight squeezes atoms in the ground, what pushes back is the combined effect of the Heisenberg uncertainty principle and the Pauli exclusion principle.

So now we can say definitively what the variety of the world is telling us: that atoms must come in many kinds, which in turn tells us there must be an edict preventing the electrons in atoms from piling up on top of each other. That edict – the Pauli exclusion principle – turns out to be an unavoidable consequence of two things: the indistinguishability of electrons and the fact that they have spin ½. It is nature's engine of difference.

The Pauli exclusion principle is not the only effect to which Pauli's name is attached. The man had a legendary effect on experimental equipment, which would invariably short-circuit, explode or simply collapse in a mangled heap whenever he was near. So bad was the 'Pauli effect' that the experimentalist Otto Stern banned Pauli from his laboratory in Hamburg and would discuss physics with him only through a closed door. However, sometimes even excluding Pauli from a laboratory offered no protection. On one occasion, when Pauli was not even on the premises, physicist James Franck was suffering a spate of equipment malfunctions in his laboratory in Göttingen. A consultation of the train timetables, however, revealed that at the time of peak laboratory mayhem, a train carrying Pauli from Zürich to Copenhagen had made a five-minute stopover at Göttingen station.

Bizarrely, Pauli himself was convinced that the Pauli effect was real. A lifelong friend of Swiss psychiatrist Carl Jung, he believed it was some kind of psychokinetic, mind-over-matter phenomenon that, although currently inexplicable, would one day come within the compass of science.

The Pauli exclusion principle has interesting philosophical implications for the hunt for the ultimate building blocks of matter. Once upon a time, these were thought to be atoms. But then the atom unexpectedly came apart, into a nucleus and a cloud of electrons. Though the main constituents of the nucleus have not been mentioned yet because they have not had a direct bearing on any discussions, it is no secret that they are the 'proton' and 'neutron'. And each of these turns out to be a composite particle too. Protons and neutrons are made of particles called 'quarks', which, incidentally, have spin ½, like electrons.

The obvious question is: have we got to the bottom of things? Or are we destined to keep breaking particles apart only to find even smaller particles inside, like a never-ending sequence of Russian dolls? Well, both electrons and quarks obey the Pauli exclusion principle, and this relies on all electrons being identical and all quarks being identical. If there is no way to distinguish one from another of its kind, then it follows that there can be no inner structure, because that would enable them to be different. The fact that electrons and quarks obey the Pauli exclusion principle is therefore a strong hint that, at long last, we have found nature's fundamental building blocks.

PART TWO

What the Everyday World Is
Telling You about Stars

We Need to Talk about Kelvin

How the fact the Sun is hot is telling you there is a source of energy a million times more concentrated than dynamite

'The great mystery is to conceive how such an enormous conflagration as the sun can be kept up. Every discovery in chemical science here leaves us completely at a loss, or rather, seems to remove farther the prospect of a probable explanation.'

John Herschel

'There's so much I don't know about astrophysics. I wish I read that book by that wheelchair guy.'

Homer Simpson (*The Simpsons Halloween Special VI*, 29 October 1995)

Walking in the park on a summer's day, you feel the soothing heat of the Sun on your face. Although the Sun is about 150 million kilometres away, it still keeps us warm. It has kept the Earth warm, in fact, for about 4.55 billion years. It may seem like a trite and self-evident observation, but actually the fact that the Sun is hot is telling us something important about the solar fuel source. Pound for pound, it must contain a million times as much energy as dynamite.

To understand why, the first thing to ask is: why is the Sun hot? The answer is surprisingly straightforward. The Sun is hot because it contains a lot of mass. It is as simple as that. Put a lot of mass in one place and its self-gravity – the gravitational pull exerted by every chunk of matter on every other – will inevitably draw all the parts closer together. The bigger the mass, the stronger the self-gravity and the more powerfully the material will be squeezed. If you have ever squeezed the air in a bicycle pump, you will know that it gets hot. The Sun is hot for exactly the same reason.

It makes little difference what the mass actually is. The Sun is mostly made of hydrogen and weighs about a billion billion billion tonnes. But put a billion billion billion tonnes of bananas or a billion billion billion tonnes of microwave ovens in one place and the end result will be the same: a glowing ball of gas pretty much as hot as the Sun. It does not make any difference what the matter is because the gravity of such a gigantic mass squeezes the material so hard that deep inside it reaches temperatures of millions of degrees. And at such dizzyingly high temperatures, atoms are slammed together so violently that many of their electrons are ripped away. The result is an electrically charged gas, or 'plasma' – an anonymous state that is the fate of all matter in such extreme conditions, regardless of whether it is hydrogen, bananas or microwave ovens.[1]

The fact that the Sun contains a lot of mass explains why it is hot – but, of course, only at this instant in time. It does not explain why it *stays* hot. After all, the Sun is continually losing heat into space, which ought to cause it to cool. It is not, though, and the implication of this is that something is replenishing the heat as fast as it is lost. But what?

On Earth, the most familiar source of heat is burning, or combustion, and as far back as 434 BC, the Greek philosopher Anaxagoras speculated that the Sun was on fire. 'The sun is a mass of fiery stone,' he said. Actually, Anaxagoras stuck his neck out a bit further than this and, with touching preciseness, speculated that 'The sun is a mass of fiery stone a little larger than Greece.'[2] Burning requires oxygen. This is often demonstrated in school science labs by placing a glass bell jar over a burning candle. As the last shreds of oxygen are consumed, the flame sputters and chokes out. Similarly, a combusting Sun requires a source of oxygen. Ignoring the small matter of where the Sun might get such a whopping source of the gas, a more pertinent question is: what exactly is the Sun burning?

Fuels which combust on Earth include wood, oil, coal, even dynamite, which liberates its heat of combustion so rapidly that it explodes instead of burning. Combustion is a chemical reaction and so involves the rearrangement of electrons around atoms, which is why wood, oil, coal and dynamite are known collectively as 'chemical fuels'.[3] So is the Sun burning a chemical fuel? Could it, for instance, be an enormous glowing lump of coal – a lump more than a million kilometres across? This may seem a daft idea. However, in the nineteenth century, when scientists first began thinking seriously about what might be powering the Sun, it was far from ridiculous. After all, they lived in an industrial society which had been made possible by the energy unleashed by burning coal.

To figure out whether the Sun is burning coal, it is necessary to know how much heat the Sun is pumping into space. Only armed with such an estimate is it possible to determine whether a lump of burning coal as big as the Sun is up to the job. The crucial measurement was made in the nineteenth century by the French physicist Claude Pouillet, and independently by the English astronomer John Herschel. The latter was the son of William Herschel, the first person to find a planet unknown to the ancients when he discovered Uranus in 1781.[4] In 1834, John sailed to Cape Town, charged with the task of extending British admiralty star charts to the southern hemisphere. With his wife, children, telescopes and belongings, he tramped across hippo-infested marshes to found an observatory on high ground, now a suburb of Cape Town called Observatory. And it was there, in 1837, on a rest day between night observations, that he successfully measured the heat output of the Sun.

Herschel and Pouillet came to pretty much the same conclusion: each year the Sun pumps out enough heat to melt a 31-metre-thick layer of ice on the Earth. It may not seem impressive, but consider that the Sun's heat streams outwards not simply in the direction of our tiny planet but in *all directions*. As a consequence, each year the heat the Sun pumps out is enough to melt a layer of ice 31 metres thick not only on the Earth but at the *distance of the Earth's orbit*. In other words, it can melt a spherical shell of ice 31 metres thick and 300 million kilometres across. Imagine an inflatable beach ball so big it swallows the Earth in its orbit around the Sun, and imagine that its surface is covered with ice to a depth of 31 metres. That is how much ice the Sun can melt in a single year, enough – to put it another way – to fill about 500 Earths.

Equipped with an estimate of the Sun's heat output, nineteenth-century scientists could seriously consider whether a coal-powered Sun was feasible. The first person to do this, in 1848, was a German doctor called Julius Mayer. He measured the heat given out by a lump of coal burning in a grate. He then scaled this up to a lump of coal the size of the Sun. The question was then: how long could such a lump of coal maintain the solar heat output measured by Herschel and Pouillet before becoming a burnt-out ember? Mayer's answer was unequivocal: no more than 5,000 years. This was a remarkably short time. It was not even enough for literal interpreters of the Bible, who believed the Earth was created on the evening of 22 October 4004 BC.[5]

Coal was ruled out as the fuel source keeping the Sun hot. But so too were all other chemical fuels, including dynamite. What, then, was powering the Sun? Mayer came up with an extraordinary possibility. The Sun was kept hot, he suggested, by meteorites which were continually raining down on it. It is a simple idea. Imagine dropping a rock from a cliff onto a pebbly beach. The rock speeds up as it falls, slamming into the beach violently. Energy comes in many forms – chemical energy, sound energy, electrical energy, and so on. And according to the law of conservation of energy, which Mayer was one of the first people to recognise, energy can neither be created nor destroyed, merely changed from one form into another. In the case of the falling rock, 'gravitational potential energy' – energy stored in the gravitational 'force field' holding everything on the Earth – is transformed into 'energy of motion'. The rock hits the beach with a sound like the crack of a gun. Whole and shattered pebbles fly through the air like shrapnel. There is even a tiny increase in the temperature of the rock and the disturbed pebbles, even the air which reverberates to the sound of the impact. Yet again the law of conservation of energy is obeyed. One form of energy – the energy of motion of the rock – is transformed into other forms of energy – the energy of motion of the scattered pebbles, sound energy, heat energy, and so on.

Heat energy is the lowest form of energy, the bottom rung of the ladder, the ultimate slag of the Universe. It is the energy of disorder, of the random, frenzied motion of microscopic atoms. Ultimately, when the sound of the rock's impact has dissipated into the air, when the pieces of pebble shrapnel have come to rest, all that will be left is heat. So when the falling rock hits the beach, in essence what is really happening is that gravitational energy is being transformed into heat energy. And this is exactly the kind of transformation Mayer had in mind when he proposed that meteorites raining down on the Sun were the source of its heat. Substitute the surface of the Sun for the beach and substitute space rocks – meteorites – for a terrestrial rock, and there you have it: Mayer's idea in a nutshell.

The 'meteoritic hypothesis' was enthusiastically embraced by William Thomson, better known as Lord Kelvin. One of the greatest scientists of the nineteenth century, Kelvin was responsible for the temperature scale used by all scientists today and also for the laying of the first transatlantic telegraph cable. He also recognised the question of what was keeping the Sun hot as one of the outstanding problems of the age. Kelvin studied the meteoritic hypothesis. Subjected to his close scrutiny, however, it fell apart. To account for the measured solar heat output, meteoritic rubble would have to be accumulating on the Sun's surface at a rate of ten metres a year. This would cause a small increase in the Sun's diameter – too small to be noticeable – so it was not the idea's Achilles heel. That lay elsewhere. Kelvin reasoned that the space rubble falling onto the Sun must be confined to the region of space closer to the Sun than the Earth. If this were not so, as the Earth orbited the Sun, it would continually sweep stuff up, changing the Earth's speed and so the length of the year. No such effect was observed. And if all the rubble falling onto the Sun was confined to the space within the Earth's orbit, there was another problem: it would have a small but significant gravitational pull. According to Kelvin's calculations, it should be enough to nudge the innermost planets, Mercury and Venus, in their orbits around the Sun. Once again, no such effect was observed.

By 1862, Kelvin abandoned the meteoritic hypothesis.

Instead, he became excited by another possibility: the idea that the Sun is kept hot because it is slowly contracting. The 'contraction hypothesis' was the brainchild of a Scottish hydrographer called John James Waterston, who had also come up with the meteorite idea, independently of Mayer, in 1853. In fact, it was a scientific paper by Waterston, not Mayer, that had got Kelvin interested in the meteorite idea. The beauty of the contraction hypothesis was its inevitability. The Sun was a giant ball of gas, and gravity was doing its utmost to shrink that ball smaller, while the force of its hot gas pushing outwards was doing its best to expand it. These two opposing forces would be in perfect, exquisite balance but for one thing: the Sun was continually losing heat to space. The lost heat robbed the gas of its ability to push outwards and defy gravity. With gravity in perpetual ascendancy, the conclusion was unavoidable: the Sun must be shrinking.

When a ball of gas contracts, it heats up. Think again of the air in a bicycle pump getting hot.[6] Another way to think of contraction is like a very slow meteor storm. However, instead of a small amount of matter falling very rapidly through the gravity of the Sun – which would happen with meteoritic heating – a large amount of matter – the entire bulk of the Sun – falls very slowly through its own gravity. Both mechanisms tap into the same ultimate source of energy – gravitational energy. And gravitational energy, as Waterston had recognised, is potentially a far bigger reservoir of energy than any chemical fuel.

Waterston's calculations showed that if the Sun were shrinking by 280 metres a year – a mere ten-millionth of its diameter and totally unobservable from Earth – this would replenish the heat continually being lost to space. The contraction idea was very promising. But it needed to be tested. Both Kelvin and his German contemporary Hermann von Helmholtz hit on a way. If the Sun were contracting today, reasoned the two men, then it must also have been contracting in the past. Once upon a time, then, the Sun must have been a gigantic cloud of gas, far bigger even than the present-day Solar System. Kelvin and Helmholtz calculated how much gravitational energy would be turned into heat as this giant cloud shrank down to the present size of the Sun. They then asked how long this heat could have kept the Sun shining at its present rate. The answer was no more than 30 million years.

A 30-million-year lifespan was a lot better than the 5,000 years for a coal-burning Sun. But, remarkably, it was still not enough. There were strong indications from the sciences of geology and biology that the Earth – and by inference the Sun, since it had to be at least as old as the Earth – was significantly older than even Kelvin and Helmholtz's estimate.

Both geologists and biologists had identified processes that had radically transformed the Earth, yet which acted so immensely slowly that they were imperceptible even in a human lifetime. In the case of geology, mountains which had once been below the sea, as indicated by fossil sea creatures on their summits, had reared miles up into the sky. In the case of biology, the tremendous variety of creatures on Earth appeared to have evolved from a simple common ancestor, morphing from one form into another by the hand of Darwinian natural selection. These transformations had shaped the Earth's surface and its flora and fauna. But to do their work they required mind-cringing, ungraspable spans of time. Not simply tens of millions of years, but hundreds of millions, maybe even billions, of years.

It was clear that an accurate estimate of the age of the Earth and Sun was needed to be sure of how long the Sun had been burning and so how much energy a solar power source required. Such an estimate would not come from geology or biology but from physics. And, paradoxically, it would involve using the unpredictability of the quantum world to create the last word in predictability – a 'clock'.

One man was pre-eminent in turning nature's quirks and foibles to his advantage, and that was Ernest Rutherford. He was the one, after all, who had the bare-faced audacity to turn the atom on itself, using one kind of atom – radium – to reveal the interior of another – gold.[7] And it was he who came up with the idea of using radioactive atoms to date rocks. The key to 'radioactive dating' was Rutherford's observation in 1900 that the radioactivity of a sample he was studying died away according to a simple law. After a certain interval of time, half the atoms remained undecayed; after the same time again, a quarter; after the same time again, one eighth, and so on. The interval was christened the 'half-life', and each radioactive substance was characterised by the length of its own half-life.

Rutherford failed to realise that this strikingly simple decay law was actually an unavoidable consequence of the irreducible randomness of the microscopic world.[8] If he had, he would have anticipated Einstein, who saw in photons the hand of a dice-throwing God. But Rutherford, though he missed out on one discovery, made another. He recognised that the radioactive decay law might provide a powerful tool for dating immensely old things.

The half-lives of different radioactive substances vary from a fleeting split second to many billions of years. In the case of uranium, which is common in some terrestrial rocks, the half-life is an enormous 4.5 billion years. Generally, the endpoint of the radioactive decay of heavy nuclei like uranium is the stable element lead. As time goes by, therefore, the amount of lead in a uranium-containing mineral grows remorselessly compared with the amount of uranium. Measuring this ratio therefore reveals how many half-lives have passed since the mineral formed and the radioactive clock started clicking. If half the uranium in a rock is left, for instance, one half-life has passed; if a quarter, two half-lives, and so on.

The American physicist Bertram Boltwood refined Rutherford's radioactive dating technique. He found that rocks from Sri Lanka were a scarcely believable 2.2 billion years old. Today, the best estimates of the age of the Earth come from the radioactive dating not of terrestrial rocks but of rocks from space. Meteorites – generally believed to be the builders' rubble left over from the formation of the Solar System – reveal that the Earth has been around for more than twice as long as even Boltwood estimated – close to 4.55 billion years.

The tremendous age of the Earth gives some inkling of the enormous energy required to keep the Sun hot. Chemical fuels are not without potency. A mere gallon of petrol contains enough energy to push a one-tonne car 40 miles. A marathon runner can go 26 miles on a plate of pasta. But a chemical fuel could keep the Sun hot for no longer than 5,000 years. The age of the Earth, which is a million times longer than this, is telling us that the fuel source of the Sun is a million times more potent than petrol or pasta.

Nuclear Energy

The first clue to what is really powering the Sun came when scientists succeeded in measuring the heat being pumped out by a radioactive substance. Although radioactivity had been discovered by French physicist Henri Becquerel in 1896, obtaining even small quantities of radioactive substances was very hard. In fact, it had taken the Herculean efforts of Marie Curie to isolate from almost a tonne of uranium ore, 'pitch-blende', the tiniest grains of the radioactive elements polo-nium and radium. But, by 1903, her husband Pierre and his colleague Albert Laborde had accumulated enough radium to measure its heat output. And what they discovered stunned them. Radium generated enough heat to boost its own weight of water from freezing point to boiling point in just 45 minutes. If this does not impress, consider that it can do the same thing in the next 45 minutes. And the next. And it can keep this up without flagging for centuries on end, thousands of years even. With a tonne of radium you could boil a tonne of water every 45 minutes – essentially for ever.

Curie and Laborde's measurement revealed the near-bot-tomless reservoir of energy inside the atomic nucleus, and it did not take long before someone latched onto this and sug-gested that the Sun was being powered by radioactivity. Here, Rutherford was guilty of a distinct failure of the imag-ination. 'The energy produced by the breaking down of the atom is a very poor kind of thing,' he declared. 'Anyone who expects a source of power from the transformation of these atoms is talking moonshine.' Rutherford was wrong about radioactivity being a poor source of power. But so too were those who believed in a radioactivity-powered Sun. Atoms of different elements give out light at characteristic wave-lengths, providing a 'fingerprint' which enables their identi-fication in a source of light. However, when astronomers examined sunlight, they were unable to detect the finger-print of radium or uranium or any other radioactive substance. Despite this, one thing was undeniable: the atomic nucleus was the seat of enormous energies. It was also just about the only candidate for powering the Sun. So if radioactivity was a non-starter for the Sun, was there another way of unleashing the energy dammed up inside the atomic nucleus?

Evidence that there might indeed be came from an unexpected direction. A shy physicist called Francis Aston made a key discovery shortly after the First World War. In a basement at the University of Cambridge, he built a machine for painstakingly measuring the masses of the atoms of different elements. Essentially, his 'mass spectrograph' measured how the trajectories of different atoms were bent by a magnetic field. If all the atoms – strictly speaking, charged atoms, or 'ions' – had the same electric charge, those bending the most had the smallest mass and those bending the least the highest. Aston made all this visible by interposing a photographic plate in the path of the flying atoms, forcing them to leave an indelible mark.

What Aston found when he measured the masses of different atoms was unexpected and astonishing. To appreciate it fully, however, it is necessary to know something about atoms – or, more precisely, the nuclei of atoms. They are built from a smaller building block. That nuclear Lego brick has the mass of a proton – a nucleus of hydrogen. (Actually, just to be awkward, nature uses two distinct building bricks, each with the mass of a proton: the proton itself, and the neutron, discovered only in 1932.) The nucleus of the lightest element, hydrogen, is made of one nuclear Lego brick; that of helium, the next heaviest, four bricks; the next, lithium, six; and so on, all the way up to uranium, which is made of 238 Lego bricks.

It stands to reason that helium, made of four bricks, must weigh four times as much as a single brick – a nucleus of hydrogen; lithium six times as much; uranium 238 times as much; and so on. But this was not what Aston found. Contrary to common-sense expectations, his mass spectrograph revealed every nucleus to weigh *less* than the sum of its constituent Lego bricks. Imagine putting ten one-kilogram bags of rice on a set of scales and it registering nine kilograms. That is the kind of bombshell Aston dropped onto the world of physics. Think of it another way. If a heavy nucleus were somehow assembled out of basic building blocks, during the process mass would vanish into thin air. But where does it go? The answer turns out to be the key to the power source of the Sun.

Common sense says mass cannot vanish. However, that is exactly what Einstein discovered in 1905. His special theory of relativity changed for ever our ideas about the nature of space and time. But it also revealed something else unexpected: mass is a form of energy. To the list of electrical energy, energy of motion and all the other myriad manifestations of energy, it is therefore necessary to add a new entity: mass-energy. The significance of this is that, according to the law of conservation of energy, energy can neither be created nor destroyed, merely transformed from one form into another. If mass is a form of energy, therefore, it follows that mass-energy can be transformed into other forms of energy. Though energy cannot disappear, mass can. But how precisely would this happen if nuclear Lego bricks were stuck together to make a larger nucleus?

The key is to focus on the force that glues everything together. Recall the rock falling onto the pebbly beach. It is the force of gravity between the Earth and the rock that pulls the two together. And when the rock hits the beach, energy is liberated – ultimately heat energy – which comes from the gravitational force field of the Earth. Well, things are pretty much the same in the case of nuclear Lego bricks coming together. There is a force that pulls them together, speeding them up. And when they hit violently, at high speed, energy is liberated – ultimately heat energy – which comes from the force field between the Lego bricks. The force between the nuclear Lego bricks is called the 'strong nuclear force' and it differs from gravity in several respects. The most significant is that it is about 10,000 trillion trillion trillion times stronger than gravity. It was not christened 'strong' for nothing.

Think again of that falling rock, the violence of its impact on the beach and the energy unleashed. Now imagine the violence and the energy released if gravity were 10,000 trillion trillion trillion times stronger. Perhaps you now appreciate the tremendous energy unleashed in the formation of a heavy nucleus out of nuclear Lego bricks. This, in a nutshell, is why the nucleus is the seat of such enormous energies.

Here, then, is why Aston's nuclei weighed less than the sum of their constituents. The enormous energy unleashed in their creation had to come from somewhere, and it came from the mass-energy of the constituent particles. Aston's mass spectrograph revealed the nuts-and-bolts reality of Einstein's theoretical prediction – that mass was a form of energy and so could be transformed into other forms of energy.

Mass-energy is special in being the most concentrated form of energy of all. The energy, E, contained in a mass, m, is expressed by arguably the most famous formula in all of science: $E = mc^2$, where c is the physicists' symbol for the speed of light. Using Einstein's formula, and the missing masses measured by Aston, it was possible to calculate the energy unleashed in the formation of a nucleus out of its constituent building blocks. The numbers were staggering. Pound for pound, the building of nuclei liberated about a million times as much energy as dynamite.

The factor of a million was highly suggestive. It was precisely the factor by which a chemical fuel like coal or dynamite fell short of being able to light up the Sun. Rutherford, who had dubbed the possibility of a power source from nuclear transformations 'moonshine', had to eat his words. 'The maintenance of solar energy no longer presents any fundamental difficulty if the internal energy of the component elements is considered to be available,' he said. 'That is, if processes of subatomic change are going on.'

Nuclear Energy and the Sun

But what processes of subatomic change could be powering the Sun? Aston's discovery certainly showed that if a nucleus were assembled from scratch out of basic nuclear Lego bricks, a dam burst of energy would be unleashed. But how likely was it that such an element build-up process was going on inside the Sun? All the building blocks seemed as likely to come together as friends just happening to converge on a street corner at the same time. A far more likely possibility was that they would arrive at the street corner one by one. Similarly, if element-building were going on inside the Sun, it was far more likely that it was going on incrementally, by the painstaking addition of one nuclear building block at a time. Actually, there was evidence that this was indeed the case. It was in Aston's data. Or rather it appeared in his data as he refined his mass spectrograph and his measurements of the mass of atomic nuclei became ever more precise.

Aston's original discovery was that the mass of every atomic nucleus was less than the sum of its constituent building blocks. In the light of Einstein's discovery, it was clear that this was because if a nucleus were to be assembled from scratch, mass-energy would go missing, transformed into other types of energy. But simply knowing how much mass-energy had gone missing in making one type of nucleus did not allow a meaningful comparison with other nuclei since, of course, some nuclei were bigger than others. A better measure to use for a comparison would be the mass-energy lost per building block. After all, the more mass-energy lost, the lighter the building blocks of a nucleus would appear to be.

Using this measure, Aston began to see a distinct pattern in his data. Nuclei of iron and nickel – pretty much average nuclei in terms of the number of building blocks they con-tained – were made of the lightest individual building blocks. Nuclei of elements with fewer building blocks than nickel and iron had heavier building blocks. And so did nuclei with more building blocks.

A graph showed the situation most clearly. Across a page, Aston plotted nuclei by increasing numbers of building blocks, ranging from hydrogen on the far left to uranium on the far right. Up the page, he plotted the weight of a nucleus's building blocks. What this showed was a valley. At the bot-tom of the valley were the nuclei of iron and nickel. High on the left-hand slope were the nuclei of small elements like helium, while high on the right-hand slope were those of big elements like uranium.

A small mass per nuclear building block is synonymous

with a lot of mass being lost in the assemblage of the nucleus out of its constituents. If a lot of mass goes missing, this can only be because those constituents are slammed together violently by a powerful force of attraction. Such a nucleus is therefore very tightly bound together and so very stable. What Aston's curve therefore shows is that nickel and iron – which have the lightest building blocks of all – are the most tightly bound, most stable nuclei in nature. For this reason, it is useful to refer to the valley as the 'valley of nuclear stability'.

All this may not at first sight appear to have anything to do with the mysterious process that the Sun is using to liberate nuclear energy. But it does.

In nature, there is a strong tendency for bodies, if at all possible, to minimise their energy. For instance, a football on the side of a valley will try to roll to the bottom, minimising its gravitational energy. Well, nuclei in Aston's valley of nuclear stability are exactly like the football. Given the opportunity, they will roll down the slopes to minimise their mass-energy. Ideally, they will roll all the way to the bottom – that is, transform themselves into nuclei of iron or nickel. However, in practice, the best they can do is roll a little way, at least in one go.

Immediately, this sheds light on why radioactivity is principally a phenomenon of big, heavy nuclei like uranium. It is because they are perched high on the right-hand slope of the valley of nuclear stability. They can reduce their mass-energy per building block by rolling down the slope – that is, disintegrating into smaller, lighter nuclei. But Aston's valley of nuclear stability also suggests another possible way of liberating nuclear energy. A nucleus perched high on the left-hand slope can also reduce its mass-energy per building block by rolling down the slope – that is, by transforming itself into a bigger, heavier nucleus. Such a process of element build-up – the complete opposite of radioactivity – would liberate surplus nuclear energy as surely as radioactive disintegration.

Aston's measurements had unexpectedly thrown up a possible nuclear process that could be powering the Sun. Could it be that, deep in the solar interior, small, light nuclei are being built up, or 'fused', into larger, heavier nuclei? The possibility was enthusiastically embraced by the English astronomer Arthur Eddington in the 1920s. Eddington was the man who had proved Einstein right, elevating him to the heights of scientific superstardom when he had measured the bending of starlight by the Sun's gravity during the total eclipse of 1919 and confirmed it was just as Einstein's theory had predicted. Asked by a physicist whether it was true he claimed to be one of only three people in the world who understood Einstein's theory, Eddington replied: 'Who is the other one?'

Eddington quickly zeroed in on the first step in the build-up process: the fusion of nuclei of the lightest element, hydrogen, into nuclei of the second lightest, helium. According to the data from Aston's mass spectrograph, in such a process a whopping 0.8 per cent of the mass would disappear, transformed into heat. More than in any other fusion process. 'I think the stars are the crucibles in which lighter atoms are compounded into more complex elements,' said Eddington.

The fusion of hydrogen into helium was promising, very promising. But there were two major problems. The first – a rather serious problem – was that the Sun did not seem to contain any hydrogen. Stamped all over its light instead was the characteristic fingerprint of iron. Taken at face value, this implied the Sun was overwhelmingly made of iron. One scientist, however, begged to differ. Her name was Cecilia Payne and she wrote arguably the most important astronomy PhD thesis of the twentieth century. Payne had a thorough understanding of quantum theory. According to the theory, every time an electron in an atom drops from one orbit to another at lower energy, the surplus energy is shed as light of a characteristic wavelength. Payne's insight was to realise that it was possible for an element to be very common yet give out little light to signal its presence. It could happen, for instance, if it was hot enough for the atoms of an element to be largely stripped of their electrons. Payne showed that this was indeed the case for hydrogen at the temperature of 5,600 degrees that is typical of the surface of the Sun.

Despite the fact that Payne's calculations showed only a tiny fraction of hydrogen retaining an electron, hydrogen on the Sun was still pumping out noticeable light. Payne realised there was only one way this could happen: if the tiny fraction of atoms was a tiny fraction of a stupendously huge number. According to her calculations, hydrogen was extraordinarily abundant on the Sun. It accounted for 90 per cent of all the atoms. The reason sunlight shows so many different wavelengths of light from iron is not that iron is common in the Sun but simply that iron atoms possess a lot of electrons – 26, to be precise – so they are almost never stripped of all their electrons. With so many electrons, and so many different orbits for those electrons to jump between, iron in the Sun emitted light at hundreds of different wavelengths.

Astronomers subsequently found that hydrogen accounted

for 90 per cent of all the atoms not just on the Sun but across the length and breadth of the Universe. They began to realise that the elements of which Earth is composed – not to mention you and me – are but minor contaminants of ordinary matter. Despite this, Payne's discovery was deeply controversial. Most astronomers of the time persisted in believing in an iron Sun. Although she had discovered the major constituents of the Universe, Payne's supervisor, the prominent American astronomer Henry Norris Russell, put pressure on her to remove any such claim. In her thesis, published in 1925, she called her result 'spurious'. Ironically, four years later, when the evidence that she was correct was overwhelming, it was Russell who received credit for the discovery. Such was the lot of a woman astronomer in the early twentieth century.[9]

To Eddington, however, Payne's work was a confirmation of what simply had to be. He believed the Sun was powered by heat energy liberated when hydrogen fused to make helium, so the Sun simply had to contain substantial amounts of hydrogen, whatever anyone said. But even if the Sun were a ball of hydrogen, there was a second serious problem with the hydrogen-fusion idea: the Sun was not nearly hot enough.

As already mentioned, the basic nuclear building blocks are glued together to form composite nuclei by the strong nuclear force, and it was pointed out that the strong force differs from gravity in several respects. One is that it is about 10,000 trillion trillion trillion times stronger than gravity. But another significant difference is that it has an incredibly short range. Until two nuclear building blocks are so close they almost touch, they feel no force of attraction whatsoever. Then – whoosh – they are caught in the microscopic equivalent of a *Star Trek* tractor beam and snapped together with an almighty thwack.

In order for hydrogen nuclei to stick together and make helium, they must therefore approach each other very closely. The strong nuclear force will then do the rest. But getting two hydrogen nuclei close enough together is easier said than done. Rutherford had deduced from his planetary model of the atom that there had to be a massive positively charged particle in the nucleus balancing the negative electric charge of the orbiting electrons – the 'proton'. The nucleus of hydrogen, the lightest element, contained a single proton. But like charges repel each other. Getting two protons close enough to feel the strong nuclear glue would mean overcoming their fierce repulsion.

In the Sun, hydrogen nuclei are in frenzied motion. The hotter it is, the faster the protons move and the more violently they slam into each other. So how hot does it have to be for hydrogen nuclei to whack into each other violently enough to overcome their mutual aversion? Eddington calculated the answer: around 10 billion degrees. So was the Sun hot enough?

Determining the temperature at the heart of the Sun without actually going there with a thermometer probably seems a tall order. But Eddington was able to estimate the temperature simply by assuming it was a ball of gas and working out how squeezed the material was at the centre. It was the old bicycle-pump effect again. Recall that the reason the Sun is hot has nothing to do with the power source of the Sun. It is hot simply because it contains a lot of mass pressing down on its interior. Eddington calculated how hot that mass makes the centre. His answer was a few tens of millions of degrees (the modern figure is about 15 million degrees). The trouble was, this was about 1,000 times too cool for the fusion of hydrogen into helium, the only known energy source that could be providing the Sun's heat. To many it would have been a serious blow. However, Eddington was convinced he was on the right track. To those who pooh-poohed his idea, saying that the Sun was not hot enough for fusion, he retorted: 'Go find a hotter place' (translated as 'Go to hell!').

Salvation came from an unexpected direction: quantum theory. Or, more specifically, the Heisenberg uncertainty principle. In Berlin, in 1929, the British physicist Robert Atkinson and the German physicist Fritz Houtermans zeroed in on the problem of how two nuclei in the Sun might get close enough to experience the strong nuclear force so they might snap together. They visualised the problem in the following way: as one nucleus approaches another more and more closely, it experiences a stronger and stronger repulsion until, finally, when it is close enough, the repulsion turns abruptly into an overwhelming force of attraction. It is like pushing a ball up a hillside that gets steeper and steeper until, finally, at the very top, there is a vertical mine shaft down which the ball plummets. The problem with a nucleus in the Sun is that it is like a ball kicked up the hill but with insufficient oomph to reach the top and fall into the mine shaft.

At least that was the situation according to the old physics. Crucially, Atkinson and Houtermans realised that things were different in quantum theory. Recall that every particle has associated with it a quantum wave and that the square of the height of the quantum wave at any place is the chance, or probability, of finding the particle at that place. The ball in the example is, therefore, not a localised thing but in some sense spread out like a humped wave on a lake. So even when it is on the hillside below the summit, its quantum wave penetrates the hillside, piercing the wall of the mine shaft drilled through the middle of the hill. Even if only the merest hint of a wave penetrates into the mine shaft, it is enough to give the ball a tiny probability of being found there – in other words, a tiny probability of it vanishing from the slope and sponta- neously appearing inside the mine shaft as if it had somehow tunnelled its way in.

This 'quantum tunnelling' is merely a consequence of the Heisenberg uncertainty principle, which is itself a conse- quence of the spread-out nature of quantum waves and the impossibility of pinning down any particle to a single loca- tion. Atkinson and Houtermans realised that it was the cru- cial missing ingredient that would allow hydrogen nuclei to fuse into nuclei of helium in the Sun at temperatures 1,000 times cooler than seemed possible. All the two physicists had to do was figure out the details of the process.

A hydrogen nucleus is made of one basic nuclear building block, and a helium nucleus of four. However, the chance of four hydrogen nuclei running into each other at the same time and sticking together is so small as to be beyond the bounds of credulity. Even worse, a nucleus with two protons does not even exist in nature – it would fly apart before it could form – so it is not even possible for two protons to col- lide and stick together. Faced with these obstacles to hydro- gen fusion, Atkinson and Houtermans had to do some lateral thinking. What they imagined was a 'proton-trapping' nucleus. This would be a relatively big nucleus which would hang about in the Sun and act as a sitting target for protons. Along would come a proton, which would collide and tunnel inside. Then another. And another. Finally, when the proton- trapping nucleus had swallowed four protons, it would suc- cumb to the nuclear equivalent of indigestion and vomit out a fully formed helium nucleus. The key thing was that, in the building of such a nucleus, 0.8 per cent of the mass-energy of the four protons would vanish, reappearing as heat energy.[10]

The question was, could such a process really generate the sunlight we feel on our face on a summer's day? Using plausible figures for the conditions in the solar interior, Atkinson and Houtermans did the calculation. To their delight, their result pretty much matched the heat output measured by Herschel and Pouillet. Two nights later, goes a story related by Houtermans, he was out with his girlfriend when she looked up at the stars and said: 'Don't they shine beautifully?' To which Houtermans replied proudly: 'I've known since yesterday why it is that they shine.'

Actually, Houtermans was jumping the gun just a little. He and Atkinson did not know the identity of the proton-trapping nucleus that was so crucial to their scheme. In fact, the two scientists were even ignorant of one of the two principal components of the nucleus. Only in 1932 would the British physicist James Chadwick discover the 'neutron', a particle with essentially the same mass as a proton but lacking any electric charge. Chadwick's discovery immediately revealed why the nucleus of the second lightest element, helium, was not twice as heavy as the first but four times as heavy; why the nucleus of the 92nd element, uranium, was not 92 times as heavy as hydrogen but 238 times as heavy; and so on. Each nucleus was padded out by neutrons. What kind of element a nucleus made depended on its complement of protons, since they were balanced by an equal number of electrons, which determined its chemistry. The mass of a nucleus, however, was determined by the total number of particles, protons plus neutrons. By convention, scientists refer to nuclei by their total mass. So the most common nucleus of uranium, element 92, is called uranium-238 because it has 92 protons plus 146 neutrons, making a total of 238 nuclear particles, or 'nucleons'. And the common nucleus of helium is called helium-4 because, in addition to two protons, it contains two neutrons.

In the light of all this, it might appear that the best way to make a helium nucleus is for two protons and two neutrons to come together and stick. Unfortunately, free neutrons self-destruct into protons and a few bits of other shrapnel after about ten minutes and so are rarely found in the interior of the Sun. With this recipe for cooking helium well and truly off nature's menu card, the only possibility remains the one in which four protons come together and stick. After Chadwick's discovery, however, it was clear that two of the protons had to metamorphose into neutrons, the nuclear equivalent of cats turning into dogs. Nobody knew exactly how this might happen inside a proton-trapping nucleus. However, it was known that in radioactive beta decay, in which a nucleus spits out a high-speed electron, a neutron in the nucleus turned spontaneously into a proton. So it was plausible that nature had a way of doing the opposite – transforming a proton into a neutron.

The man who picked up the baton from Houtermans and Atkinson was Hans Bethe, a Jewish nuclear physicist who had been forced to leave Germany when Hitler came to power in 1933. After a conference on the stellar energy problem held in the US in 1938, Bethe suddenly realised that he possessed enough knowledge of the properties of different nuclei finally to identify the elusive nuclear reactions that were powering the Sun. According to a story related by the Ukrainian-American physicist George Gamow – the conference organiser and a man renowned for his colourful tales –

on the train back to New York Bethe announced: 'It should not be so difficult after all to find the reaction which would just fit our old sun. I must surely be able to figure it out before dinner.' And, on his napkin, Bethe proceeded to figure out the chain of nuclear reactions that would result in the fusion of hydrogen to helium.

Bethe deduced that the proton-trapping nucleus had to be carbon, and the chain of nuclear reactions he scrawled across his napkin became known as the carbon–nitrogen–oxygen, or CNO, cycle, because nitrogen and oxygen were involved as well. Coincidentally, the CNO cycle was discovered at the same time in Germany by Carl-Friedrich von Weizsäcker, the son of the second-highest official in Hitler's foreign ministry. So finally, after hundreds of years of speculation, the power source of the Sun – hydrogen fusion by the CNO cycle – was identified.

Not quite.

Bethe and von Weizsäcker's chain of nuclear reactions was certainly one way to fuse hydrogen into helium and liberate the enormous energy packed into the atomic nucleus. However, there was another possibility. In 1932, the American chemist Harold Urey had found a heavy form of hydrogen. Unlike ordinary hydrogen, hydrogen-1, which contained a single proton, heavy hydrogen, hydrogen-2, contained a proton and a neutron. The discovery of hydrogen-2, christened 'deuterium', prompted a rethink of the sunlight-generating nuclear reactions in the Sun.

The non-existence of a stable nucleus containing two protons blocked the path to building helium simply by adding protons one at a time. This was, of course, why it had been necessary to postulate the existence of a proton-trapping nucleus. But the existence of deuterium opened up the possibility of two protons colliding in the Sun and making deuterium, a nucleus which was stable. Was this possible? If it was, then deuterium might act like a mini-proton-trapping nucleus, with further proton hits providing a simpler route to forging helium than the CNO cycle.

Bethe found that the deuterium route was indeed possible. And working with his student Charles Critchfield, he worked out the details. Von Weizsäcker came up with the same process independently in Germany. As with the CNO cycle, tunnelling was at the heart of the 'proton–proton chain'. A proton tunnelled into a deuterium nucleus to make a nucleus of helium-3, a light isotope of helium, and then a helium-3 nucleus actually tunnelled into another helium-3 nucleus. This created the desired nucleus of helium-4, with two protons left over to start the process again.

The question was, which process – the CNO cycle or the proton–proton chain – was powering the Sun? Since a proton approaching a proton would experience less repulsion than a proton approaching a more highly charged carbon nucleus, the proton–proton chain could operate at lower temperatures than the CNO cycle. Everything therefore depended on the precise temperature at the heart of the Sun. Gradually, it became clear that at the temperature at the heart of the Sun, the proton–proton chain would win out over the CNO cycle. It would be far more efficient. In a hotter, more massive star than the Sun, on the other hand, it would be the CNO cycle. The proton–proton chain of nuclear reactions was the elusive source of sunlight.

The proton–proton chain is actually a three-step process. But the crucial step is the first one – the creation of deuterium from two protons. Such a conversion requires the existence of a new force of nature known as the weak nuclear force. The force has been dubbed 'weak' because it truly is weak compared with the strong nuclear force. But the most important thing to know is that in the quantum world, where forces are caused by the exchange of force-carrying particles, weak is synonymous with slow. It is because the first step is slow that it is so crucial. Just as the slowest member of a team of cyclists in a Tour de France time trial governs the team's speed, the conversion of a proton into a neutron in the creation of deuterium controls the speed of hydrogen fusion in the Sun.

And here the word 'slow' really means slow. The average proton in the Sun spends an astonishing 10 billion years or so flying about before it collides and sticks with another proton and changes into a neutron to make deuterium. This means that to generate sunlight, the Sun is using just about the most inefficient nuclear reaction imaginable. Believe it or not, your stomach generates heat at a faster rate than an equivalent volume of the solar interior. If you wonder how the Sun can be so inefficient and yet pump out so much heat, the answer is simply that there is a lot of the Sun. And like all big objects, its surface area – through which heat can escape – is relatively small compared with its volume. So heat tends to be bottled up inside, boosting the temperature of the Sun.

The fact that it takes a pair of protons flying about the interior of the Sun 10 billion years on average to get together and make deuterium, the first step in the proton–proton chain, determines the solar lifetime. The Sun consumes its hydrogen fuel at such a parsimonious rate that it will take about 10 billion years before it runs out. Since the Sun has already been burning for close on 5 billion years, there are about 5 billion years to go.

Now it turns out that the first step in the proton–proton

chain, in which a proton hits another proton, changes into a neutron and sticks, is only possible because of a fluke. The fluke has to do with the strength of the strong nuclear force. If it were just 1 or 2 per cent stronger than it is, it would be strong enough to overcome the mutual aversion of two protons and glue them together to make a nucleus with two protons. Since, in the quantum world, strong is synonymous with fast, this reaction would be extremely fast. In fact, all of the Sun's hydrogen fuel would be used up in less than one second, exploding the Sun in the process.

We owe our very existence to this fluke. If the strong force were a few per cent stronger, the Sun would not exist. Because the strong force has the strength it has, the Sun can exist for 10 billion years, enabling vast tracts of time for the evolution of complex living organisms such as us. Consequently, the fact that the Sun is shining in the sky today is not only telling us that there is an energy source in nature a million times more concentrated than dynamite. It is telling us about an amazing fluke in the strength of nature's fundamental forces on which our very existence depends.

5

You, Me and the Spectacularly Unlikely Triple-Alpha Process

*How your very existence is telling you that our
Universe may not be the only one*

'As we look out into the Universe and identify the many accidents of physics and astronomy that have worked together to our benefit, it almost seems as if the Universe must have known that we were coming.'

Freeman Dyson

'And I say to any man or woman,
Let your soul stand cool and composed before a million universes.'
Walt Whitman ('Song of Myself')

Look around you. The Earth is teeming with life. Nobody knows how it got started. But one thing is for sure. Life as we know it could not have begun without the element carbon. Atoms of carbon have a unique ability to join up with others of their own kind, creating a bewildering array of complex molecules. In our bodies, carbon-based 'biomolecules' perform myriad essential tasks, from metabolising the food we eat to responding to the light falling on the retina of our eye to encoding the information of inheritance in deoxyribonucleic acid, or DNA. We are carbon-based bipeds whose existence depends on carbon being common. After hydrogen, helium and oxygen, it is the fourth most abundant element in the Universe. The abundance of carbon actually tells us something rather interesting. It is telling us about a series of spectacularly unlikely coincidences in the properties of a handful of atomic nuclei. Not only are they responsible for your existence, but – even more than this – they strongly hint that our Universe is but one among an infinity of other universes floating like bubbles in an unimaginably huge 'multiverse'.

It is an extraordinary conclusion to draw from the mere fact of our existence, but the logic turns out to be pretty inescapable. The first thing to realise is that the elements, including carbon, were not placed in the Universe on Day One by the Creator. Instead, the Universe started out with only the simplest of nuclear building blocks – protons and neutrons – and, subsequently, these have glued themselves together to form the nuclei of the 92 naturally occurring elements.

The evidence that the elements have been *made* – built up a brick at a time – is actually rather subtle. One of the most important clues comes from the abundances of different elements throughout the Universe. These can be estimated in many ways. One is by analysing the composition of rocks from the Earth's crust and of meteorites from space. Such measurements were first carried out by the Swiss-Norwegian chemist Victor Goldschmidt in 1936. The abundance of elements can also be measured by examining the characteristic fingerprint they create in the light coming from stars, a technique used to effect by Cecilia Payne when she surprised everyone by discovering that the Sun is made overwhelmingly of the lightest elements, hydrogen and helium. It is interesting here to recall the words of the French philosopher Auguste Comte, who, in 1835, compiled a list of things he believed could never be achieved by science: 'Never by any means shall we be able to study the chemical composition or mineralogical structure of the stars.' Within 25 years, the

German physicist Gustav Kirchoff showed that elements such as sodium, when heated in a flame, give out light at characteristic wavelengths and that such 'spectral fingerprints' can be used to identify different elements in the light from the Sun and stars. Comte was saved the ignominy of having to eat his words by having died before the discovery.

The analysis of the composition of the stars, Earth rocks and meteorites threw up a striking result. All across the cosmos, the elements appeared to be present in roughly the same relative proportions. As the American physicist Richard Feynman said: 'The most remarkable discovery in all of astronomy is that the stars are made of atoms of the same kind as those on the Earth.' Less striking, but equally important, is the pattern that is visible in the 'universal' abundances. In general, the heavier an element, the rarer it is in nature. In fact, so steep is the decline in the abundance of elements that uranium, element 92, is a billion times less common than element 11, sodium. It is easiest to see this on a piece of graph paper. If elements are plotted along the page, getting ever more massive from left to right, and their abundances plotted up the page, the result is a mountainside. From light elements on the left of the page, the mountainside plunges precipitously downwards to heavy elements like uranium on the far right.

Some elements, however, buck the trend of rapid decline in abundance with increasing mass. They stand out as significantly more common than their neighbours on the mountainside. For instance, there are hummocks at carbon, nitrogen and oxygen, and also around iron and its immediate neighbours. In addition, some elements are distinctly less common than their neighbours. For example, there are hollows on the mountainside at lithium, beryllium and boron.

Why are some elements more common than expected and others less so? A strong clue can be found in a surprising place: Aston's valley of nuclear stability.

Recall that, in Aston's valley, nuclei with the lowest mass per nucleon – iron and nickel – are found at the bottom, and on the slopes rising on either side are nuclei with more and more mass per nucleon. Well, it turns out this simple picture does not tell the whole truth. As Aston refined his mass spectrograph and was able to measure the masses of nuclei ever more precisely, he found that the sides of the valley were not entirely smooth. There are little hillocks at locations where a nucleus has more mass per nucleon than its immediate neighbours, and there are little potholes at places where a nucleus has less mass per nucleon. The peculiar thing is that bumps on the abundance mountainside coincide exactly with potholes on Aston's valley of nuclear stability, and hollows on the abundance mountainside coincide with hillocks on Aston's valley of nuclear stability. The conclusion is unavoidable: there must be a connection. How common an element is must depend on the detailed properties of its atomic nucleus. It is a strong hint that nuclear processes are behind the creation of the elements – that the elements have been *made*.

Imagine that, high up on the slopes of a valley, someone lets loose a load of footballs. As they roll down the slope towards the valley bottom, they avoid the bumps on the slope and get trapped in the potholes. The correspondence between the cosmic abundances of the element and Aston's curve shows that something like this must really have happened in nature. Nuclei must have been released high on the left-hand side of Aston's valley of nuclear stability. They then 'rolled' down the slope towards the valley bottom, avoiding the bumps and lodging in the potholes. A nucleus high on the left-hand side of Aston's valley of nuclear stability is a small, light nucleus. One that rolls down towards the valley bottom is therefore a light nucleus getting heavier and heavier as it accrues nuclear building blocks, one at a time. It is a light nucleus being built up into a heavier nucleus.

But if the elements were made – as all the evidence indicates – where were they made? The clue is in the temperature required for element building. Bigger, heavier nuclei have a bigger electrical charge than smaller, lighter nuclei. They therefore repel each other more fiercely, which means higher temperatures are required to smash them together hard enough to stick. The hottest places in the Universe appear to be stars like the Sun. Unfortunately, calculations by the English astronomer Arthur Eddington in 1925 showed that stars could not be the cosmic furnaces in which the elements were forged. According to him, the rotation of the Sun caused the material inside to circulate endlessly, and this endless circulation would keep the stuff of the Sun thoroughly mixed. So when hydrogen fused into helium to make sunlight, the helium ash would be mixed evenly throughout the star. The trouble was, this would continually dilute the hydrogen fuel of the Sun. As time passed, the Sun would gradually cool and go out. This was the opposite of what was required for an element-building furnace.

In the US, George Gamow was aware of Eddington's calculations. They spurred him to begin looking about for another furnace that might be hot enough to forge the elements. He soon found one: the fireball of the Big Bang. In 1929, the American astronomer Edwin Hubble, working at Mount Wilson Observatory in southern California, had discovered that the galaxies – the building blocks of the

Universe of which our own Milky Way is but one among billions – are flying apart from each other like pieces of cosmic shrapnel in the aftermath of a titanic explosion. We live in an expanding universe. And because it is expanding, an unavoidable conclusion is that it must have been smaller in the past. In fact, if the expansion is imagined running backwards like a movie in reverse, we come to a time when all of creation was compressed into an infinitesimally tiny volume. This was the moment of the Universe's birth in the Big Bang, currently believed to have occurred 13.7 billion years ago.

Gamow picked up the idea of a Big Bang and ran with it. If the Universe had been smaller in the past, he reasoned, then it must also have been hotter (that old air-in-the-bicycle-pump effect again). The Big Bang must have been a 'hot' Big Bang. And if the Big Bang was hot, could it not have been the furnace in which the elements were forged out of some simple basic ingredient? The trouble was that time was in short supply. By the age of ten minutes, the Universe had expanded and cooled to such an extent that element-building processes were effectively choked off. It was actually a double whammy. By this time, the fireball of the Big Bang was no longer dense enough for nuclei to encounter each other frequently; and, when they did run into each other, they were moving too slowly to overcome their mutual repulsion. Gamow was not easily put off. Ten minutes simply had to be enough. 'The elements were cooked in less time than it takes to cook a dish of duck and roast potatoes,' he maintained.

Gamow's optimism was misplaced, however. There was a more serious obstacle to element building than the limited time available. In nature there is no stable nucleus containing either five or eight basic building blocks. It means it is nigh on impossible to build any nuclei heavier than helium. After all, if a single nuclear building block – either a proton or a neutron – were to collide with and stick to a helium-4 nucleus, it would make a nucleus of mass 5. But no stable nucleus of this mass exists. And if two helium nuclei were to collide and stick, they would make a nucleus of mass 8. Once again, no such stable nucleus exists. There is no way to get beyond helium. Disappointingly for Gamow, the Big Bang could not have been the furnace in which the elements of nature were forged.[1]

Enter the British astronomer Fred Hoyle. Stars, to Hoyle, seemed a far more attractive furnace in which to forge the elements. After all, they remain dense and hot for millions, if not billions, of years, rather than the mere ten minutes of the Big Bang's fireball. Since so much time was available, there was the possibility of relatively infrequent nuclear processes working their magic. All it would take would be one such rare process to leapfrog the troublesome mass-5 and mass-8 barrier and the road to the forging of heavy elements would be wide open. The trouble was that any such process would undoubtedly require very high temperatures. However, Eddington had shown that as stars fused hydrogen into helium, they gradually cooled and snuffed out. Hoyle was not discouraged. Out in space there were stars which were huge and bloated and pumped out as much as 10,000 times as much heat as the Sun. The existence of such 'red giants' – a prime example of which is Betelgeuse, burning brilliantly in the constellation of Orion – is evidence that there must be a way for stars to avoid the ignominious end envisaged by Eddington.

One way a star might do this, Hoyle realised, was if heavier elements became more common in its core than in its outer envelope. Since this would make the core denser than its surroundings, the powerful self-gravity of the core would crush it, heating it up. The temperature might easily soar to 100 million degrees, exactly what the doctor ordered for fusing together light nuclei to make heavier ones. At the same time, the intensely hot core of the star would pump out prodigious amounts of heat into the surrounding stellar envelope, causing it to puff up to giant proportions. And, as the material of the star inflated, it would cool until it merely glowed a dull red. This was a recipe for a red giant. It convinced Hoyle he was on the right track.

The trouble, of course, was that Eddington's calculations showed that the material of a star was always thoroughly mixed together. But Hoyle was not to be put off. With astronomer Ray Lyttleton, he concocted a way of circumventing Eddington's stellar show-stopper and forming a star with a super-dense, super-hot core loitering at the heart of a puffed-up and bloated envelope. It required the pair to postulate the existence of dense, cold clouds of hydrogen gas drifting about the Galaxy. Nobody knew whether such clouds existed. But if they did, Hoyle and Lyttleton pointed out, a star circling the centre of the Galaxy would inevitably plough through them, gathering about itself a thick mantle of hydrogen gas. Its interior – since it would be a mix of helium and hydrogen – would then be denser than its exterior. It was the recipe for creating a red giant with a super-dense, super-hot core.

Hoyle's idea was ingenious, but it was unnecessary. Eddington was the pre-eminent astrophysicist of his day, but he suddenly realised he had made a stupid numerical error in his calculations. He was correct that the rotation of the Sun caused the material in its interior to circulate endlessly.

However, he was wrong about the speed of that circulation. It was enormously more sluggish than he had estimated. In fact, it was so slow that it could not possibly keep the material of the solar interior blended. Without such mixing going on, the core would become ever more rich in helium as hydrogen burned. It would get denser, shrink and heat up. A star like a red giant, it turned out, was the natural and unavoidable fate of a star like the Sun.[2]

Hoyle's hunch was right. Stars could, after all, get hot enough for element-building. But there remained the problem of the mass-5 and mass-8 barriers, which Gamow had discovered blocked the way to cooking up the elements in the furnace of the Big Bang. Hoyle looked around for a rare nuclear process that might leapfrog the barriers. And he found one. It involved not two helium nuclei but three. Could it be possible that, deep in the helium-rich cores of red giants, helium nuclei – alpha particles – came together in threes? If they stuck, the result would be a nucleus of carbon-12, which would neatly bypass the mass-8 hurdle.

This 'triple-alpha' process had actually already been considered by the American physicist Ed Salpeter in 1952. Salpeter realised immediately that the chance of three helium nuclei coming together simultaneously was so unlikely as to be effectively impossible – think of three blindfolded footballers blundering about a football pitch and all running into each other simultaneously at the corner flag. Instead, Salpeter focused his attention on two helium nuclei coming together. It might seem like a non-starter since, of course, gluing two helium nuclei together would make a nucleus of mass-8, which is unstable. But what Salpeter realised was that although such a nucleus – beryllium-8 – was unstable, it was not totally unstable. It hung about for a split second before falling apart. And, crucially, for that split second it was a sitting duck for a third helium nucleus.

The triple-alpha process, instead of requiring a ridiculously unlikely three-body encounter between helium nuclei, might be accomplished with more mundane two-body processes. Salpeter envisaged two steps. First, a pair of helium nuclei would collide and stick to make beryllium-8. Then, before the beryllium-8 nucleus had a chance to disintegrate, it would be struck by another helium nucleus to make a nucleus of carbon-12.

Salpeter's two-stage triple-alpha process was far more likely than its single-stage version. Unfortunately, it was still not likely enough. When Salpeter carried out the calculations for the core of a red giant, he found the triple-alpha process could convert no more than a tiny fraction of a star's helium into carbon. It was too inefficient. It was a dead end.

Hoyle was aware of Salpeter's failure. However, he was unwilling to abandon the triple-alpha process because, quite frankly, it was the only game in town. Could there be a way to speed things up, he wondered? As he turned the problem over and over in his head, it struck him that there was indeed a way to boost the efficiency of the triple-alpha process. The trouble was, it was an awfully long shot.

Imagine a child on a swing. Say they are moving forwards and backwards once every five seconds. If you push the swing every three or every seven seconds, you will fail to boost the size of the arcs of the swing, and pretty soon you will have a protesting child on your hands as the swing comes to an erratic halt. Push the swing every five seconds, however, and it goes ever higher. Physicists say that the swing has a 'natural frequency' of one swing every five seconds. And it is a characteristic of any oscillating system like a swing that when the driving force – in this case, you pushing – matches its natural frequency, energy is transferred most efficiently. The oscillating system is said to be 'in resonance', or to 'resonate'.

Now consider an atomic nucleus – specifically a nucleus of carbon-12. Imagine it as a bag containing a dozen nuclear building blocks. In reality, there is no such bag. However, the strong nuclear force which binds the building blocks effectively confines them to a small volume exactly as if they are in a bag. Now within the bag the nuclear building blocks jostle back and forth ceaselessly. Actually, the jostling is not entirely random. There is evidence that within a nucleus, the nucleons orbit in tight 'shells' reminiscent of the shells of orbiting electrons. But the key thing is that the whole bag has certain frequencies at which its contents naturally oscillate or vibrate.

Frequency is synonymous with energy, with sluggish, low-frequency vibrations containing little energy and violent, high-frequency vibrations a lot of energy. So each internal vibration of a carbon-12 nucleus corresponds to a particular vibrational energy. And energy was the key to Hoyle's big idea for boosting the speed of the triple-alpha process. If three helium nuclei – or, equivalently, a beryllium-8 nucleus and a helium nucleus – collided and their total energy was *exactly* that of one of carbon-12's natural vibrations, there would be a resonance. It would be like pushing the swing at its natural frequency. Only in this case what would be pushed would be the speed of the nuclear reaction that glued the components together to make carbon-12.

Of course, the nuclear reaction would be resonant only if carbon-12 happened to have an 'energy state' that exactly matched the combined energy of motion of three helium nuclei at the typical 100-million-degree temperature in the heart of a red giant.[3] Hoyle put in the numbers and calculated the energy. It was 7.65 megaelectronvolts (MeV). Precisely what an MeV is is not important. Suffice to say, it is a unit physicists find convenient to express the energy of an atomic nucleus. The important thing is that if carbon-12 has an energy level at precisely 7.65 MeV, the nuclear reaction to make carbon-12 from three helium nuclei is resonant. Hoyle calculated how much carbon-12 would be forged in the heart of a red giant, assuming the 7.65 MeV energy state existed. It was appreciable. The speeded-up triple-alpha process worked. The mass-5 and mass-8 barriers were bypassed. The road to the building of all heavy elements was wide open. Everything depended on carbon-12 having a vibration energy at precisely 7.65 MeV. The question was, did it?

As luck would have it, in the spring of 1953, Hoyle was on a sabbatical from Cambridge University at the California Institute of Technology in Pasadena, where there was an experimental nuclear physics group. In fact, it had even dabbled in 'nuclear astrophysics'. Its measurement of the speed of the crucial nuclear reactions in the CNO cycle were critical in showing that the CNO cycle could be the power source only of stars significantly more massive than the Sun. On arrival at Caltech, Hoyle wasted no time in going to the Kellogg Radiation Laboratory to see the group's leader, Willy Fowler, and asking his question. Could carbon-12 have an energy level at 7.65 MeV?

He might as well have asked whether fairies were orchestrating the nuclear reactions in the heart of the Sun. Fowler thought he was in the presence of a lunatic. No theorist had ever been able to predict the precise energy of a compound nucleus. The mathematics was just too complicated.

Although physicists rarely admit it, the only physical system whose behaviour they can predict with certainty is the two-body one: the Moon moving under the influence of the Earth's gravity; an electron in a hydrogen atom orbiting in the electromagnetic grip of a proton. When it comes to three or more bodies, theorists are flummoxed. And carbon-12, with a dozen particles buzzing about in its nucleus like a tight knot of bees, is a 'many-body' system. It is totally beyond the power of theorists to predict its properties precisely. But that was exactly what Hoyle – a bespectacled young astronomer from England – was claiming to have done.

What made Hoyle's prediction even more ridiculous was the crackpot logic behind it. 'I exist and I am made of carbon, so the 7.65 MeV energy level of carbon-12 must exist,' was the gist of it. In all of his research career, Fowler had never heard anything so extraordinary. A conclusion drawn from the observational fact that humans exist. An 'anthropic' argument. Biology determining physics. Scientific reasoning turned on its head.

It was highly probable Hoyle was wrong. On the other hand, Fowler adhered to the experimenter's maxim: never close your mind to the unexpected. He rounded up the members of his small research group, who listened politely while Hoyle repeated his extraordinary argument for the existence of the 7.65 MeV state of carbon-12. Was there any possibility, Hoyle asked, that the experiments to date could have somehow missed it? Much of the technical discussion that followed went way over Hoyle's head. Eventually, however, Fowler's colleagues reached a consensus. If the 7.65 MeV state of carbon-12 had certain very special properties, yes, it was just possible that experiments might have missed it. The team decided to rejig their equipment and take a look.

For ten days, as the experiment proceeded, Hoyle was on tenterhooks. Each afternoon, he crept down into the bowels of the Kellogg Lab, the benevolent gift of a cornflake magnate, where Fowler's colleague Ward Whaling and his team beavered away amid a jungle of power cables, transformers and diving-bell-like chambers in which atomic nuclei were fired at each other. And each afternoon, he crept back up again into the painfully bright Californian sunshine, relieved that his idea had survived one more day without being blown out of the water but anxious for the next day, and the next. On the tenth day, Hoyle was met by Whaling, who pumped Hoyle's hand and gushed his congratulations. The experiment had succeeded. Hoyle's prediction had been borne out. Unbelievably, there was an energy state of carbon-12 within a whisker of 7.65 MeV.

It was the most amazing result that Fowler had ever witnessed. He had not really believed that Hoyle's outrageous prediction would be proved right. But it had – spectacularly. Like some omniscient god, Hoyle had peered into the heart of nature and spied something that mere mortals – or, at least, mere nuclear physicists – had been unable to see. He had maintained that the 7.65 MeV energy state of carbon-12 must exist because, if it did not, neither could human beings. To this day, Hoyle is the only person to have made a successful prediction from an anthropic argument in advance of an experiment.

Despite the spectacular triumph, however, Hoyle was not out of the woods. Once a carbon-12 nucleus formed inside a red giant, it was a sitting duck, just waiting to be struck by another helium nucleus. The result would be a nucleus of oxygen-16. All the good wrought by the triple-alpha process would be undone. Although carbon would be made, it would promptly be transformed into oxygen. The Universe would be carbon-free.

For carbon to be forged, it was necessary for a carbon-12 nucleus to vibrate at a very special energy equal to the combined energy of three helium nuclei at the typical temperature at the core of a red giant. Hoyle now realised that in order for some carbon to survive and not be transformed into oxygen, it was also necessary for oxygen-16 *not* to vibrate at a particular energy. Specifically, it must not have an energy equal to the combined energy of a carbon-12 nucleus and a helium nucleus at the temperature of a red giant. If it did, there would be a resonance and all carbon-12 would promptly be transformed into oxygen-16.

As part of its work on the CNO cycle, Fowler's team had already measured the properties of the oxygen-16 nucleus. Hoyle pored over the data. There was a heart-stopping moment when he saw that oxygen-16 had an energy state very close to the energy to be avoided at all costs. But when he looked in detail, he found to his relief that the energy state was just out of range. Oxygen would indeed be made inside stars but, fortunately for human beings, not at the expense of carbon.

When Hoyle had time to think about what he had discovered, he began to marvel at the nuclear coincidence on which our existence so crucially depended. Beryllium-8 was unstable, but not so unstable that the triple-alpha process was impossible. Carbon-12 had an energy level at just the right place so that the triple-alpha process would be resonant and thus make appreciable quantities of carbon. And oxygen-16 had no energy level at a particular place, so that not all the carbon-12 would be transformed into oxygen-16. Without these three conditions being satisfied, the Universe would contain no elements heavier than carbon, or, alternatively, heavy elements but no carbon. Instead, everything was finely balanced to produce a Universe with roughly the same quantities of carbon and oxygen, both of which were essential for life.

Hoyle wondered what to make of this, and he came up with two logical possibilities. One was that there is a God who has fine-tuned the properties of the nuclei of beryllium-8, carbon-12 and oxygen-16 so we can be here. The problem with this option is that it is not a scientific one. A striking characteristic of science is that you get out more than you put in. A scientific explanation – often distilled into a formula or equation – is always simpler and more compact than the observations it summarises. If God fine-tuned things, the explanation – a complex supreme being – is as complex, if not more complex, than the thing for which an explanation is being sought. You get out less than you put in – the antithesis of science. The other problem with the God hypothesis is that one of the most striking things about the Universe is that it appears to be running perfectly well according to the known laws of physics, without any supernatural input.

But if a Creator did not fine-tune the energy levels of beryllium-8, carbon-12 and oxygen-16, what is the explanation for these unlikely nuclear coincidences? Hoyle came up with a stunning possibility. Perhaps our Universe is not the only one. Perhaps there are many universes, each with different laws of physics. In most the laws do not conspire to create nuclear coincidences for the creation of carbon, so there is no life. It is no surprise, then, that we find ourselves in a universe with the nuclear coincidences necessary for life.

How could we not be? It is amazing, topsy-turvy logic. But, to Hoyle, it was the only thing that made sense. Incredibly, the fact that we exist as carbon-based beings may not simply be telling us about nuclear coincidences deep inside stars. It may be telling us that out there, in other spaces or other dimensions, there are a large number – perhaps an infinity – of other universes.

6

The 4.5-Billion-Degree Furnace

How the fact iron is common on Earth is telling us there must be a furnace out in space at a temperature of at least 4.5 billion degrees

'What is precious is never to forget
The essential delight of blood drawn from ageless springs
Breaking through rocks in worlds before our earth.'
Stephen Spender

'Come quickly, I am tasting stars.'
Dom Pérignon (at the moment
of his discovery of champagne)

You walk down the concrete canyon of a city and iron is all around you. It is in buildings which could never climb so high without reinforced skeletons of steel. It is in the railings that border the parks. It is in the cars passing by, the planes flying high above, the train rumbling over the bridge up ahead. Even the blood coursing through your veins and arteries contains iron, the essential component of haemoglobin, which ferries oxygen ceaselessly around your body. Iron is the most abundant element on Earth. It composes 5 per cent of the Earth's crust and makes up most of the Earth's core. Perhaps you have never wondered about the ubiquity of iron on Earth. But actually it is telling us something important: that somewhere out in space there must be a furnace at a staggeringly high temperature of 4.5 billion degrees.

The straightforward conclusion to draw from the ubiquity of

iron on Earth is that the element-building nuclear reactions inside stars must continue adding nuclear Lego bricks all the way up to iron. Iron is just about the most tightly bound, most stable nucleus in all of nature. It sits at the very bottom of Aston's valley of nuclear stability. Given the chance, every light nucleus will roll down to the bottom of that valley and transform itself into iron. It can do this by gobbling one light nucleus after another. However, this is not quite as easy as a football rolling down a hill because the electric charge carried by nuclei causes them to repel each other fiercely. Only if this repulsion can be overcome can nuclei stick together and grow bigger. In practice, this requires the nuclei to be slammed together violently, something which happens only at ultra-high temperature inside stars, when nuclei are flying about at ultra-high speed. The bigger nuclei grow, the bigger their electrical charge and the higher the temperature necessary to overcome their mutual aversion. Since iron is a moderately big nucleus – constructed from 56 nuclear building blocks and carrying 26 times the charge of a hydrogen nucleus – its synthesis requires an extremely high temperature. Precisely how high, however, depends on the exact sequence of nuclear reactions involved in building it.

Recall that, in the heart of a red giant star, where the temperature pushes 100 million degrees, helium-4 nuclei slam into each other in threes to forge nuclei of carbon-12.[1] If a carbon-12 nucleus, made by this triple-alpha process, is then struck by another helium-4 nucleus, it is transformed into oxygen-16. And this is just the beginning of a long series of nuclear reactions in which a nucleus grows by capturing one helium-4 nucleus, or alpha particle, after another. This 'alpha process' can keep going just as long as the temperature inside the star keeps climbing. An oxygen-16 nucleus, by accreting a helium-4 nucleus, is transformed into neon-20. Neon-20, by doing the same, is morphed into magnesium-24. Magnesium-24 into silicon-28. And so on.[2]

The forging of silicon-28 is of critical importance to the fate of a star. Once a star reaches this stage in its life, the next steps in the alpha process – collectively known as 'silicon burning' – race ahead at breakneck speed. Helium nucleus after helium nucleus is added in rapid, rat-a-tat-tat succession. Within a single day, the star arrives at the very brink of making iron – and, as it turns out, total catastrophe. Actually, the addition of the last alpha particle in the long chain actually makes nickel-56, not iron-56. But this nucleus is radioactively unstable and disintegrates quickly into cobalt-56, which then decays into iron-56. Because the last step in the silicon-burning process involves the biggest, most highly charged nuclei, it is this step that determines the temperature needed to forge iron in stars. That temperature is about 4.5 billion degrees. So this is what the ubiquity of iron on Earth is telling us: that somewhere out in the Universe there must be stars whose internal fires burn 300 times hotter than the heart of the Sun.

Attaining such an enormous temperature is a very tall order, and in practice the only stars that can get this hot are monsters more than ten times as massive as the Sun. But the existence of such stars, and such a blisteringly hot furnace, is not enough in itself to explain why iron is so common on Earth. After all, the iron that is forged must somehow get out into space, and it would appear to be securely locked up inside a star.

Exactly how locked-up is evident from the way in which the interiors of stars evolve to the silicon-burning, iron-forging stage. Recall that when a star turns hydrogen into helium in its core, the core becomes helium-rich. Because it is heavier than the exterior of the star, it shrinks and, in the process, heats up. This creates the conditions for helium to fuse to make carbon. So the star now has a helium-burning core surrounded by a hydrogen-burning shell. But as the star turns helium into carbon, the core becomes carbon-rich and heavier than the rest of the star, so it in turn shrinks and heats up. This creates the conditions for carbon to burn. Now the star has a carbon-burning core surrounded by a helium-burning shell surrounded by a hydrogen-burning shell. By now the pattern should be clear. As a star evolves, building up heavier and heavier nuclei, its interior develops a structure reminiscent of an onion. The heaviest elements are made in the core, the next lightest elements in a shell wrapped around this, the next lightest in a shell about that, and so on. By the time a star is turning silicon into iron in its heart, it consists of a large number of element-building onion skins. From the outside in, there are shells of hydrogen, helium, carbon, oxygen, neon . . . and, finally, iron.

The problem, it would appear, with element-building in stars is that all the elements painstakingly built up over the lifetime of the star are locked away inside. And of all the locked-up elements, none is more locked up than iron, sequestered at the very heart of the onion-like star. The fact that iron is ubiquitous on Earth is therefore not simply telling us that there is a furnace at 4.5 billion degrees somewhere out in space. It is telling us there must also be a way of getting the iron from that furnace out into space so it can enrich the gas clouds floating there – gas clouds which provide the birth material for new stars and planets. That way was discovered at the height of the Second World War by the British astronomer Fred Hoyle.

Like many scientists, Hoyle was assigned war work. In his case it was the development of radar for the ships of the British navy. At the end of 1944, as part of this work, he was sent to a radar conference in Washington DC. When it was over, it had been arranged by the British Embassy that he would fly across the country to San Diego to see a radar unit at the US naval headquarters. Once in southern California he knew he was only a few hundred miles from Mount Wilson Observatory, which boasted the biggest telescope in the world – the 100-inch Hooker telescope. So, in a spare weekend in his itinerary, he took the train north to Los Angeles.

It was at Mount Wilson that Edwin Hubble had made the most important cosmological discovery of the twentieth century: that the Universe is expanding, its constituent galaxies flying apart like pieces of cosmic shrapnel in the aftermath of a titanic explosion – the Big Bang. When Hoyle arrived at Mount Wilson Observatory's headquarters in Pasadena, most of the astronomers were absent, engaged like their British counterparts in war work. A notable exception was Walter Baade, a German émigré who at the outbreak of war had been classified as an enemy alien and so excluded from military service. This had allowed him to work at the 100-inch, taking advantage of the air-raid-blackened night skies over Los Angeles to see deep into the Universe.[3] After Hoyle had spent a weekend on Mount Wilson at the 100-inch, it was Baade who was sent to pick him up and drive him back down to Pasadena. During the journey, the conversation happened to turn to the latest developments in astronomy and, in particular, exploding stars. Hoyle said he thought they were dull, but Baade quickly persuaded him they were anything but.

Baade's particular interest was in 'supernovae', which he had recognised as a distinct class in the 1930s. The key to this recognition was Hubble's discovery of the true nature of galaxies, which had preceded his discovery that they were fleeing from each other. In 1924, Hubble had used the 100-inch telescope to pick out individual stars in the Great Nebula in Andromeda, thus proving it was not a gaseous nebula within our Milky Way, as many astronomers believed, but an entirely separate island universe of stars – a galaxy – way beyond the edge of the Milky Way. In fact, using a type of star known as a Cepheid variable as a 'distance indicator', Hubble was able to determine the distance to Andromeda. Despite being the nearest galaxy, it was so far away that light took millions of years to travel to Earth.

This had implications for exploding stars, or 'novae', which Baade had been observing with his Swiss colleague, Fritz Zwicky.[4] They had spotted detonations in galaxies. It was suddenly clear that these must be enormously more luminous than the ones observed in the Milky Way. Baade and Zwicky suggested that there must be two distinct classes of exploding stars in the Universe. 'Supernovae' were a million times more powerful than ordinary novae, often outshining an entire galaxy of 100 billion stars.

What could cause such a tremendous conflagration? Where did all the energy come from? Hoyle pondered this question as he travelled to Montreal to catch a flight back to England. He very quickly realised there was only one source of energy big enough: gravitational energy. If a massive star ran out of fuel in its core, it would no longer be able to generate heat to oppose gravity. The core would be crushed mercilessly. In the rapidly shrinking central regions of the star, the temperature would rise to unimaginable levels, causing nuclei to break apart into their constituent protons and neutrons. This catastrophic collapse would even crush all the empty space out of matter, squeezing protons and electrons together to make neutrons. In a matter of seconds, the middle of the star would be transformed into a hard ball of neutrons no bigger than Mount Everest.[5] Hoyle did not know how this implosion might be turned into the explosion of a supernova. All he knew was that in the shrinkage of the heart of the star down to a 'neutron core', enough energy would be converted into heat to power a supernova. It was gravitational energy again, the same source Kelvin had mistakenly believed was powering the Sun.

Hoyle's flight home from Montreal was delayed because of bad weather. During his stay in the city, he was surprised to bump into friends from back in Cambridge. They were nuclear physicists and Hoyle knew they had been recruited to a shadowy project called Tube Alloys, which he was fairly sure was a front for the British programme to build a nuclear bomb. The idea of building a bomb had been in the air since late 1938, when, in Berlin, Otto Hahn and Fritz Strassman had announced that a nucleus of uranium, when struck by a neutron, split into two. Since this 'fission' spat out more neutrons, which could split more uranium nuclei, it raised the spectre of a runaway 'chain reaction' that might unleash an unstoppable tidal wave of nuclear energy. All that was needed to make a bomb of unimaginable ferocity was a large enough lump of uranium.

In 1939, Niels Bohr and John Wheeler had determined that such a runaway chain reaction could be triggered only in a rare type of uranium. Hoyle had assumed that Tube Alloys' goal was to accumulate a lump of this uranium-235. When this was done, it would send its personnel to North America to test a bomb, far away from the prying eyes of the Germans. However, uranium-235 was chemically identical to the common form of uranium, and separating it out was a formidable task which Hoyle imagined was likely to take years. It seemed impossible to him that Tube Alloys had succeeded so soon. The mystery then was why its personnel were in Montreal.

The only explanation Hoyle could think of was that there must be another, faster route to the bomb, and this must be close to a test explosion. And as a matter of fact, Bohr and Wheeler had indeed found that there was another nucleus, apart from uranium-235, that would undergo fission. Plutonium was not known in nature but might be made artificially from uranium in a nuclear 'pile', or reactor. It was the perfect material for a bomb because it was a distinct element and so could easily be separated from uranium.

As Hoyle marked time in Montreal, waiting for the weather to clear, he recalled a rumour he had heard that somewhere in the south-western US a bomb team had been assembled from some of the greatest scientific minds in the free world.[6] It seemed odd to him that such a formidable team was needed. He had thought triggering an explosive release of nuclear energy merely required taking two lumps of fissionable material and slamming them together. Once above their 'critical mass', a runaway chain reaction would automatically ensue. The existence of the American bomb team indicated to Hoyle that clearly this would not work for plutonium. Somewhere along the plutonium route to a bomb there must be an obstacle. It must explain why Britain had chosen the uranium-235 route that Hoyle had thought was so hard.

So what was the plutonium problem? The only thing

Hoyle could think was that when two pieces of the man-made element were slammed together, so much heat was generated by the fissions that it pushed apart the pieces before a runaway chain reaction could take hold, causing a fizzle rather than an explosion. If so, the pieces would have to be forced together. This could be done by surrounding the plutonium with a spherical shell of conventional explosives and imploding it. But Hoyle knew that creating a perfectly spherical blast wave was fantastically hard. This, he finally deduced as he waited for his plane in Montreal, must be the obstacle standing in the way of a plutonium bomb.

At this point, thoughts that had been circling in Hoyle's head since his meeting with Baade began to come together. It was the word 'implosion' that did it. Implosion, he believed, drove the explosion of a supernova. And implosion was evidently what would drive the runaway nuclear reactions of a plutonium bomb. Hoyle put two and two together. Could it be, he wondered, that implosion in a supernova drove the nuclear reactions that built up elements? Could it be that supernovae were the furnaces where iron and all the other elements in our bodies were made?

When the weather cleared, Hoyle flew in a Liberator bomber back to Scotland, high above the U-boat-infested waters of the North Atlantic. Finally, back in Cambridge, he continued to think about the inferno in a supernova as the perfect furnace for building up elements. He realised immediately that in the hell of a stellar implosion, the temperature would rise so incredibly high that every conceivable nuclear reaction would be possible. Nuclei would also be squeezed so close together that those nuclear reactions would occur at a breakneck rate. Such super-fast nuclear reactions create a state called 'nuclear statistical equilibrium', in which there is a perfect balance between the processes of creation and destruction. Every nuclear reaction and its opposite runs at exactly the same rate, so each element is built up exactly as fast as it is broken apart. Imagine if water is fed into a tank at exactly the rate it leaks out. Despite the toing and froing, the water level stays precisely the same. Similarly, in nuclear statistical equilibrium the abundance of each element remains unchanged, or 'freezes out'. Contrary to all expectations, the bewilderingly complex orgy of nuclear reactions has a simple outcome. And crucially, Hoyle realised, that outcome is predictable.

The abundances would depend on only two things: the temperature in the shrinking stellar core, which determined the average energy of motion of the nuclei; and the mass differences between the nuclei, which determined which nuclei would be preferentially made by nuclear reactions. All Hoyle needed was those mass differences. As luck would have it, he bumped into Otto Frisch, the Austrian physicist who in 1939 had first alerted the British government to the danger that the Germans might build an atomic bomb. Frisch had recently returned from the US, where he had been working on the bomb project. He had what Hoyle wanted. From a drawer in his desk, he pulled out a table of nuclear masses which had been compiled by a German physicist called Josef Mattauch.

Hoyle knew that iron and elements of similar mass formed an unusually broad hummock on the abundance mountain slope. It had been christened the 'iron peak'. It rose from gently sloping foothills in the vicinity of scandium, up through titanium, vanadium, chromium and magnesium to the summit at iron-56, the most abundant of the iron-group elements. On the far side of the summit, the peak plunged steeply down through cobalt and nickel to foothills in the neighbourhood of copper and zinc. Using Mattauch's table of nuclear masses, Hoyle calculated the frozen-out abundances expected for nuclear statistical equilibrium inside an imploding star. To his amazement and delight he saw that, for a temperature of about 5 billion degrees, those abundances precisely matched the shape of the iron-peak abundances.

It was an epochal moment in the quest to discover the origin of the elements. Hoyle had found the unmistakable fingerprint of element-creation in nature. The discovery convinced him beyond any doubt that stars could indeed become hot enough to forge the elements in our bodies, and that other places, like Gamow's Big Bang furnace, were not that furnace.[7] It would later set him on course to figuring out how the triple-alpha process might make carbon, opening the way to building up all heavy elements inside stars. The big bonus of Hoyle showing that element-building must go on inside supernovae was that a supernova exploded. It would scatter the products of its furnace to the winds of space. There they would later be incorporated into newborn stars and planets, and life like ours.

The supernova story actually goes like this. A massive, highly evolved star eventually develops an onion-like interior, with heavier and heavier elements in shells closer and closer to its core. Finally, it undergoes silicon burning, which creates a core of iron. But this is very bad news for the star. Iron is at the bottom of Aston's valley of nuclear stability, so when an iron-56 nucleus is struck by a helium nucleus, the result is a nucleus with more energy per nucleon, not less.[8] Instead of mass-energy being turned into heat – which has been the case with all the element-building processes in the life of the star up to this point – heat must be turned into mass-energy. The only place that heat energy can come from is the core of the star itself. In other words, heat is sucked vampire-like from the heart of the star. Deprived of heat, the core will be unable to oppose the gravity trying to crush it. It will shrink catastrophically.

And it is in the maelstrom of the imploding core that nuclear reactions of nuclear statistical equilibrium go on. Those nuclear reactions forge the iron-peak elements we find on Earth. They range from titanium to chromium to copper and zinc to iron itself. They are the elements that have made civilisation possible, and for that we have the furnaces of supernovae to thank. The elements ejected into space by supernovae are a combination of nuclei the star has painstakingly built up over its lifetime *and* the nuclei forged in the hell of the explosion itself. Nobody said nature was simple.

In 1987, the picture of element-building inside an exploding star was put to the test when the first supernova visible to the naked eye in 400 years burst into the sky. It detonated in the Large Magellanic Cloud, a satellite galaxy of our Milky Way, and historic photographic plates showed that the precursor star was a massive sun called Sanduleak −69° 202. The star reached the state where it was burning iron, which sucked the heat out of its core and caused it to implode. Eventually, the temperature in its heart got so high that nuclei came apart into neutrons, which welded together to make a neutron core – a tiny, super-hard ball the size of Mount Everest. The in-falling star literally bounced off the hard surface of this neutron star. This is how implosion was turned into explosion, creating supernova 1987A.

Before the explosion, as the core of the imploding star

shrank, the rising temperature not only triggered an orgy of nuclear reactions that created the elements of the iron peak, it also triggered nuclear reactions in the onion-skin layers around the core – layers made mostly of carbon-12, oxygen-16, neon-20 and silicon-28. These became so blisteringly hot that nuclei in them rolled all the way down to the bottom of the valley of nuclear stability. As pointed out before, the most stable nucleus in nature is not iron-56 but nickel-56, which shares the same number of nuclear building blocks. Consequently, it was into nickel-56 that nuclei in the cooler onion-skin layers surrounding the supernova core transformed themselves.[9] Nickel-56 spits out a high-energy photon known as a gamma ray, decaying into cobalt-56, with a half-life of six days. In turn, cobalt-56 spits out another gamma-ray photon, decaying into iron-56, with a half-life of 77 days.

The significance of this is that the gamma rays can be detected by gamma-ray observatories in orbit around the Earth. The best supernova 1987A data actually came from the second decay, when cobalt-56 turns itself into iron-56. Astronomers detected these gamma rays. Not only did they have the expected energy but they faded away with a characteristic timescale of 77 days. It was strong enough evidence that the picture of element-building in supernovae is correct.

But there is more.

As the debris of Sanduleak −69° 202 expanded, by rights it should have quickly cooled and faded. However, there was nickel-56 in the debris, and the gamma rays from its decay reheated the debris,[10] making the debris glow with visible light. In other words, the gamma rays were responsible for the very light by which astronomers saw the supernova. And that light died away on a characteristic timescale of 77 days. Nickel was turning into iron before the astronomers' eyes. They were witnessing the clear signature of the formation of iron-56 in a supernova explosion.

The picture that emerged was awe-inspiring. The iron on Earth and in our blood comes from massive stars which detonated as supernovae before the Sun was born.[11] Their debris intermingled with the gas and dust floating between the stars. Out of that gas and dust, 4.55 billion years ago, there congealed the Sun and planets. It is a story which is unique in connecting the very big and far away – stars – to the very small and close to home – the atoms of which you and I are composed. Hold up your hand in front of you. You are made of star stuff. As the American astronomer Allan Sandage has remarked: 'All humans are brothers. We came from the same supernova.' This is what the abundance of iron on Earth is telling you.

PART THREE

What the Everyday World Is Telling You about the Universe

The Unutterable Feebleness of Starlight

*How darkness at night appears to be telling us there was
a beginning to time but is actually telling us
something quite different*

'If the stars are other suns having the same nature as our sun, why
do not these suns collectively outdistance our sun in brilliance?'
  Johannes Kepler (*Conversations with the Starry Messenger*, 1610)

'The only way in which we could comprehend the blackness our
telescopes find in innumerable directions would be by supposing
the distance of the invisible background so immense that no ray
has yet been able to reach us.'
  Edgar Allan Poe ('Eureka', 1848)

*It is a crystal-clear night far away from the lights of any town
or city. The stars are shining like diamonds. There are so many
stars that they distract from the most striking feature of the
night sky: it is black. Overwhelmingly black. It may seem like a
trite observation. However, it is telling us something important
about the Universe. The overwhelming majority of astronomers believe that it is telling us that the Universe has not existed
for ever; that there was an instant when it came into being;
that, in common with you and me and every creature on Earth,
the Universe was born. But actually the world's astronomers
are dead wrong. The darkness of the sky at night is telling us
something entirely different.*

The person who first realised that such a commonplace
observation of the sky might have something to tell us about
the cosmos was the German astronomer Johannes Kepler,

imperial mathematician to the emperor of the Holy Roman Empire. In 1610, he received a copy of Galileo's best-seller *The Starry Messenger*, in which the Italian scientist documented the astronomical discoveries he had made with the newly invented telescope. They included mountains on the Moon and the four 'Galilean' moons of Jupiter. Kepler was so inspired by the book that he dashed off a letter to Galileo, which was later published as a short book. In *Conversations with the Starry Messenger*, Kepler not only underlined the importance of Galileo's work but pointed out something that nobody else appeared to have noticed: the darkness of the sky at night is deeply surprising.

Most people, if asked why the sky is dark at night, would say that it is because there is no Sun and starlight is much weaker than sunlight. It takes a genius to realise that the reason it is black at midnight is far from obvious and may actually have something profound to say about the Universe.

Kepler's reasoning was straightforward. If the Universe is infinite in extent so that its stars march on for ever, then between the bright stars in the night sky we should see more distant, fainter stars, and between them, stars even more distant and even more faint. It was like looking into a dense forest. Between the trunks of some nearby trees you see the trunks of more distant trees and, between them, the trunks of trees even further away. The view that confronts you is therefore of a solid wall of trees. Similarly, claimed Kepler, when we look out into the Universe, we should see a solid wall of stars.

It is possible to be more precise than this. Imagine the Earth is surrounded by spherical shells of space rather like the concentric skins of an onion. The further away a shell, the fainter the stars it contains. On the other hand, the further away the shell, the bigger it is, containing more stars. Well, the increase in the number of stars should exactly compensate for the stars getting fainter.[1] In other words, every onion-shell of stars should contribute exactly the same amount of light to the terrestrial night sky. But this is disastrous. If the Universe goes on for ever, there are an infinite number of such shells. Add up the light coming from all of them and the answer is an infinite amount. Far from being dark at night, the sky should be infinitely bright.

Infinity – a number bigger than any other – is merely an abstract mathematical concept. Nothing in the real world is infinite in size. The conclusion that the night sky should be infinitely bright must therefore be wrong. There must be a flaw somewhere in the logic. And there is. Although the stars appear to be dimensionless pinpricks, in reality they are other suns, shrunken to mere specks by their immense distance. Each is a tiny disc – too small to see with the most powerful telescopes – but a disc nonetheless. Consequently, the discs of nearby stars obscure those of the faraway ones, just as nearby trees in a forest hide the faraway ones. This means the night sky should be papered entirely by the discs of stars. Although not infinitely bright, it should be as bright as the surface of a typical star.

Kepler believed the Sun was a typical star. Consequently, he concluded that the night sky should be as bright as the surface of the Sun. We know today that the Sun is not an average star. It is considerably more luminous than most. About 70 per cent of stars in the solar neighbourhood are 'red dwarfs', cool suns reminiscent of softly glowing embers. However, this hardly changes Kepler's conclusion. Rather than being as bright as the surface of the Sun, the sky at night should be glowing blood red from horizon to horizon. 'In the midst of this inferno of intense light', said the Anglo-American cosmologist Edward Harrison, 'life should cease in seconds, the atmosphere and oceans boil away in minutes, and the Earth turn to vapour in hours.'

Thankfully, the sky is not as bright as the surface of a typical star. It is about a trillion trillion trillion times fainter. The paradox that the night sky is dark when, logically, it should be bright ought to be called Kepler's paradox. However, because it was popularised by a distinguished German astronomer called Heinrich Olbers in the early nineteenth century, it has instead become known as Olbers' paradox.

When a prediction clashes with a cast-iron observation, clearly it is the prediction that is at fault. More than likely the assumptions that went into making the prediction need re-examining. Kepler's most obvious assumption was that the Universe goes on for ever. If this is not true, then the paradox can go away. After all, there will be only a limited number of onion shells of stars contributing their starlight to the Earth's night sky. It is easy to imagine the sky being filled with so little starlight as to appear black. This was actually Kepler's solution to the dark-sky paradox. He abhorred the idea of an infinite Universe. It terrified him. It was monstrous. He therefore concluded, with some relief, that the Universe must be finite in extent.

If Kepler was right, the cosmos is not like an endless forest; it is akin to a localised clump of trees bounded at the rear by a dark wall. Because the clump is so small and sparse, we can see the dark wall behind. This is the blackness between the stars.

As a matter of fact, in the twentieth century astronomers did indeed discover that the Universe is finite – or at least the portion of the Universe from which we receive starlight. Recall Edwin Hubble's 1929 discovery that the Universe is expanding, its constituent galaxies flying apart like pieces of cosmic shrapnel. If the expansion is imagined to run backwards, like a movie in reverse, there comes a time when all of creation is squeezed into the tiniest of tiny volumes. This was the beginning of time, the moment of the Universe's birth, the Big Bang. According to the best current estimates, space, time, matter and energy exploded into being in the fireball of the Big Bang about 13.7 billion years ago.

The size of the Universe – or at least its effective size – is inextricably linked to its age. This is because light, though fast, is not infinitely fast, so it takes time for it to cross space.[2] An interval of 13.7 billion years may seem an unimaginably huge tract of time, but it is simply not long enough for light, crawling snail-like across the vastness of space, to have made it to Earth yet from the most distant reaches of the Universe. Consequently, the only celestial objects we can see are those whose light has taken less than 13.7 billion years to reach us. Imagine them occupying a bubble of space – the 'observable universe' – centred on the Earth.

The observable universe is bounded by the 'cosmic light horizon'. This is pretty much like the horizon at sea. We know there is more of the sea over the horizon. Similarly, we know there is more of the Universe over the cosmic light horizon, only its light has not got here yet. It is still on its way.

A light year is the distance light travels in a year.[3] So an obvious conclusion to draw is that the distance to the cosmic light horizon must be 13.7 billion light years. However, this is incorrect since the Universe, in its first split second of existence, is believed to have undergone a brief, faster-than-light epoch of expansion. Because of this 'inflation', the distance to the light horizon is not 13.7 billion but about 42 billion light years.[4]

Of course, the Universe may be infinite in extent. In fact, in the inflationary picture it is effectively infinite. However, the combination of the finite age of the Universe and the finite speed of light reduces the volume of space from which we can receive light to a bubble 84 billion light years across. This cuts the amount of light arriving on Earth dramatically.[5]

Remarkably, the first person to realise that the night sky might be black because there were stars too far away for their light to have got to us was Edgar Allan Poe. In his imaginative essay 'Eureka', published in 1848, he wrote:

> Were the succession of stars endless, then the background of the sky would present us a uniform luminosity since there could be absolutely no point, in all that blackness, at which would not exist a star. The only way in which we could comprehend the blackness our telescopes find in innumerable directions would be by supposing the distance of the invisible background so immense that no ray has yet been able to reach us.

It would seem, then, that the evidence that the Universe has a finite age – that it was born in a Big Bang – stares us in the face every night. In fact, it has been staring people in the face since the dawn of human history. Only nobody realised. Nobody guessed the true cosmic significance of dark sky at night.

It is a wonderful story. It is a neat resolution of a 400-year-old mystery. It is a story that 99 per cent of the world's professional astronomers will trot out if you ask them. Unfortunately, it is not true.

In concluding that in an infinite Universe the night sky should be as bright as the surface of a typical star, Kepler made a hidden assumption: that stars are actually up to the job of filling the Universe with light. After all, for the night sky on Earth to appear bright, the empty space between the stars must be filled to the brim with starlight. All that can be deduced from the fact that the sky is dark at night, therefore, is that for some reason the stars have not yet succeeded in filling up the Universe with their light.

Light is constantly being pumped into empty space by countless trillions upon trillions of stars. But, as Douglas Adams pointed out in *The Hitch Hiker's Guide to the Galaxy*: 'Space is big. You just won't believe how vastly, hugely, mind-bogglingly big it is.' Consequently, it would take a very long time for space to fill with starlight like a bath filling to the brim with water. Exactly how long was calculated in 1964 by Edward Harrison. His answer was a whopping 100,000,000,000,000,000,000,000,000 years.[6] This is much longer than the 13.7 billion years that the Universe has been in existence.

The darkness of the sky at night is therefore telling us not that the Universe was born – which would be a truly remarkable thing – but merely that it must be younger than the light fill-up time of 100,000,000,000,000,000,000,000,000 years. This is not a very useful constraint on the age of the Universe. Nevertheless, it was obtained at no cost, deduced from the most mundane observation of the sky at night.

Can we therefore conclude that we have come on the cosmic stage rather too early? And that if we were to wait patiently for 100,000,000,000,000,000,000,000,000 years, the night sky would indeed be as bright as the surface of the average star? Of course, after such a vast interval of time, the

Earth would be long gone, most likely swallowed by the Sun, which itself would have winked out and died. But that is a minor detail. Imagine that, for the sake of argument, even after 100,000,000,000,000,000,000,000,000 years there is still a convenient vantage point from which to observe the Universe. Would the whole of space be glowing as brightly as the surface of a typical star? The answer is no. The reason is that not only would the Sun be long dead, *all the stars* would be long dead. They simply do not burn long enough.

This was something Kepler failed to realise. The idea that an energy source is needed to heat something and that eventually all energy sources are exhausted is a relatively recent one. At least, it was not appreciated before the nineteenth century. So far, the Sun has burned for about 5 billion years but it will run out of energy within another 5 billion years. Red dwarfs, which are much less massive than the Sun, live longer. Their fires are cooler, so they use up energy at a miserly rate and may survive for 100 billion years or more.[7] But even this span of time pales into insignificance compared with the 100,000,000,000,000,000,000,000,000 years needed by the stars to fill the Universe to the brim with light.

It turns out that Kepler's – or Olbers' – paradox was never actually a paradox after all. The night sky could never be as bright as the surface of the average star because that would require the stars to fill up space with starlight, and they simply do not contain enough energy to create the required starlight. The night-sky paradox is telling us that either the Universe is younger than the time needed to fill it with light or that the stars have insufficient energy to make a bright sky. It turns out the latter is true. The sky is dark at night because there is not enough energy in the Universe. It is as simple as that. The stars are too feeble by far.

Since mass is a form of energy, it is possible to ask how much matter would have to be converted into light to fill the Universe with starlight. The answer is 10 billion times more matter than exists in the Universe. But stars do not convert 100 per cent of their mass into starlight, only about 0.1 per cent. Consequently, the stars fall short of being able to fill the Universe with starlight by a factor of about 10 trillion.

So if the darkness of the sky at night is telling us not that the Universe was born but merely that the stars are unutterably feeble, why discuss it at such length? Because it is a fascinating historical story, for one thing. Because if you know the true explanation for darkness at night, you will know something that 99 per cent of the world's professional astronomers do not know. Because, sometimes, knowing why a wrong-headed argument is wrong-headed – and this one baffled the world's best brains for 400 years – can itself illuminate important things. And because, actually, there does exist everyday evidence that tells us the Universe was born – this chapter will not be short-changing you – it is just not the darkness of the sky at night.

Tune your TV between stations. Some of the static, or 'snow', on the screen is caused by the microscopic jitter of electrons in the circuits of the TV. Some is from radio waves picked up from buildings, trees, the sky, keys turning in the ignition of cars, passing taxis, and so on. But about 1 per cent of the static is from radio waves which have come directly from the Big Bang itself. Before being intercepted by your TV aerial, they had been travelling for 13.7 billion years across space. And the last thing they touched before your aerial was the blisteringly hot fireball at the birth of the Universe.

The afterglow of the Big Bang fireball is in the air around us, which is why, at this very moment, your TV aerial is picking it up. Remarkably, 99.9 per cent of the light in the Universe is tied up in this 'cosmic background radiation', with only 0.1 per cent in the form of starlight. In fact, every sugar-cube-sized volume of space is currently being criss-crossed by about 300 photons from the Big Bang. Which makes it all the more incredible that the afterglow of the Big Bang was not discovered until 1965 – and then entirely by accident.

The credit goes to two young radio astronomers who were employed by Bell Telephone Laboratories in Holmdel, New Jersey. Though it may seem odd that a commercial company employed two astronomers, there was method in Bell Labs' madness. By the early 1960s, the company had seen the future, and that future involved bouncing telephone signals around the world via satellites high in Earth orbit. This meant Bell Labs would have to develop the ability to detect faint radio signals from minuscule objects in the sky – which is where the radio astronomers came in. They too were in the business of trying to register faint sources of radio waves in the sky – for instance, distant galaxies. If Bell Labs employed some radio astronomers, went the reasoning, it stood to benefit from their specialist expertise.

In turn, Arno Penzias and Robert Wilson were attracted by the prospect of using a unique radio antenna. Bell Labs' engineers had been using a giant 'microwave horn' at Holmdel to transmit and receive radio signals at a 'micro-wave' wavelength of 7.35 centimetres to and from the *Telstar* communication satellite. But when that project was finished, they abandoned the horn. Penzias and Wilson jumped at the chance to use it for astronomy. They were to be sorely disappointed, however.

Penzias and Wilson intended to use the Holmdel antenna – an ice-cream-cone-shaped horn the size of a railway carriage – to try and detect faint radio waves coming from cold hydrogen gas which they believed to be floating in space in the outer regions of our Milky Way. Since they expected the signal from such gas to be extremely weak, before they could make any astronomical observations they first needed to account for all the spurious sources of radio waves in the neighbourhood of their antenna so they could subtract them from what the antenna picked up. Everything warm – you, me, trees, buildings, the sky, and so on – emits radio waves from jiggling electrons.

Pretty soon, however, Penzias and Wilson hit a problem. After subtracting every source of spurious radio waves from their signal, they found that their horn was still registering a persistent radio hiss. It was present wherever they pointed the horn at the sky. And it was exactly the emission that would be expected from a body at an ultra-chilly –270°C – three degrees above 'absolute zero', the lowest temperature possible.

At first, Penzias and Wilson thought the anomalous hiss might be coming from New York City, which was just over the horizon from Holmdel. But when they pointed the horn away from New York, the hiss did not disappear. Then they thought it might be coming from electrons injected high into the atmosphere by a nuclear test the previous year. But as the months passed, the static did not fade as expected. They thought it might be coming from an unknown source within the Solar System. But as the Earth travelled around the Sun, changing its position relative to other bodies in the Solar System, the static did not change. They even thought the hiss might be the microwave glow from pigeon drop-pings. Two pigeons were nesting deep inside the horn – a cosy, warm place during the harsh New Jersey winter – and they had coated the interior with what radio engineers quaintly referred to as a 'white dielectric material'. The material, better known as pigeon shit, glowed gently with microwaves. Penzias and Wilson ousted the pigeons, posting them in the company mail to another Bell Telephones site, then climbed into the horn with stiff brooms to sweep away the offending droppings. But after their hard work, they found to their dismay that the persistent static was still there.

What Penzias and Wilson had stumbled on – and this became clear only when they learnt of a nearby team at Princeton University, which by a bizarre coincidence was actually looking for it – was the leftover heat of the Big Bang. It had been the physicist George Gamow who had first realised that if the Universe had once been small, it must also have been hot. And if it had been hot, the heat must still be around today. After all, it had absolutely nowhere to go. It was bottled up in the Universe, which, by definition, is all there is. In 1948, Gamow's students, Ralph Alpher and Robert Herman, realised that the expansion of the Universe over the past billions of years would by now have cooled down the heat of the Big Bang. At only a few degrees above absolute zero, it would appear not as visible light but as invisible microwaves. Microwaves – radio waves with a wavelength ranging from a few centimetres to a few tens of centimetres – were given out by all manner of celestial objects. But Alpher and Herman realised that the afterglow of the Big Bang would be distinguishable from all other sources. For a start, being cosmic in origin, it would be coming equally from all directions in the sky.[8]

What Penzias and Wilson had picked up was the leftover heat from the Big Bang – the afterglow of creation. It is so ubiquitous that if your eyes could see microwaves rather than visible light, at night the whole sky would appear to be glowing white from horizon to horizon. It would be like being inside a giant light bulb.[9]

This ought to ring a bell. Remember that for the night sky to be bright, the stars must fill empty space to the brim with starlight. In practice, however, this means them shining for 100,000,000,000,000,000,000,000 years, which is impossible, since they do not have the energy reserves necessary. But what is impossible for stars turns out to be possible for the afterglow of the Big Bang. Or at least once upon a time it was possible. In the fireball of the Big Bang, there was indeed sufficient energy for light to fill all of space. Today, since the Universe has become so much bigger, diluting and cooling the light of the fireball, there is no longer enough energy around. Yes, if you had microwave eyes, you would see all of space glowing – but they would have to be very sensitive microwave eyes.

But your TV *is* a very sensitive microwave receiver. Turn it on again and tune it between the stations. That static is telling you that the Universe has not existed for ever, that it was born in a Big Bang, that there was a beginning to time.

8

The Bang Before the Big One

How the fact that the Universe has been essentially non-quantum for most of its history is telling us it must once have undergone a burst of super-fast expansion

'If you can look into the seeds of time,
And say which grain will grow, and which will not,
speak then to me.'

Shakespeare (*Macbeth*)

'Many and strange are the universes that drift like bubbles in the foam upon the River of Time.'

Arthur C. Clarke ('The Wall of Darkness')

In the everyday world, a cause always precedes an effect. You brake at a pedestrian crossing because moments earlier someone stepped out into the road; you get drenched by the rain because a cloud burst overhead. In the everyday world, things happen with absolute certainty. The Sun will rise tomorrow; Mars will in six months' time be exactly where Newton's laws predict it will be, so NASA can be confident its robot space probes will reach the Red Planet. In the everyday world, a Martian space probe – or a football kicked through the air – follows a single, well-defined path towards its destination, not half a dozen separate paths simultaneously.

These statements may appear so ridiculously self-evident as to be unworthy even of mention. But actually the way the everyday world behaves is very surprising. After all, the fundamental theory that orchestrates our Universe is quantum theory. And a central characteristic of quantum theory is that things happen with no prior cause. There is no telling whether a particular photon will bounce off a window or go right through, whether a radioactively unstable atomic nucleus will sit quietly for the next billion years or self-destruct in the next millisecond. What actually happens is irreducibly unpredictable, in common with everything else that happens in the quantum world. And when a photon or any other denizen of the microscopic world flies through space, it does not follow a single path but, in some sense, all possible paths simultaneously.

The most striking feature of our Universe – one hardly ever remarked upon, even by physicists – is that it behaves in a largely non-quantum manner. And not just at this particular moment in time. Because of the finite speed of light, telescopes act as time machines, revealing past epochs of the Universe. And what those time machines tell us is that for the majority of cosmic history, the Universe has been behaving in a pretty non-quantum way.

It is a paradox. We live in a quantum universe that largely looks un-quantum. This is a profound observation about reality. And, remarkably, it may be telling us something about the very birth of the Universe. What is that thing? In its earliest moments, the Universe must have undergone a burst of super-fast explosive expansion.

To understand how it is possible to deduce something so precise about the primordial Universe from the fact that we live in a non-quantum world it is first necessary to say something about the theory of the beginning of the Universe. Our most successful theory of physics, as already mentioned, is quantum theory. Because it describes the microscopic world of atoms and their constituents, it may not be obvious that it has anything to say about the large-scale universe. However, the Universe has expanded from a highly compressed state in the Big Bang. In its earliest moments, therefore, the Universe was indeed smaller than an atom. So if we want to understand the earliest moments of creation, we need a quantum theory of the Universe – a theory of quantum cosmology.

In practice, this means having a quantum theory of gravity, since gravity orchestrates the behaviour of large masses such as the Universe.[1] Such a theory is often dubbed the holy grail of physics since it would unite the theory of the very small – quantum theory – with the theory of the very big – Einstein's theory of gravity, the general theory of relativity. Unfortunately, although physicists have been able to describe the three non-gravitational forces in quantum terms, they have failed, despite many decades of trying, to do the same for gravity.[2]

Not having a quantum theory of gravity, and so a quantum theory of cosmology, would appear to be a fatal handicap in any speculation about the beginning of time. Remarkably, however, it is still possible to say something about the early Universe. Recall that it is a characteristic of quantum theory that when a particle travels between A and B, it can do so by travelling along every conceivable path, each of which has a certain 'probability' of being taken. It is likely, therefore, that a successful theory of quantum cosmology will view the history of the Universe not as a single thread but as a whole bunch of strands all bundled together. The challenge is then to discover why our Universe has followed its own particular history rather than any other.

One way to do this is to use observations of our present-day Universe to rule out a host of possible histories in the hope that the only history we will be left with is one like our own. And the most striking observation about our Universe, as already pointed out, is that for most of its history it has been non-quantum. Consequently, we can discard the myriad possible cosmic histories in which the Universe stays smaller than an atom and dominated by quantum effects such as quantum unpredictability. This naturally leaves only universes that grow big.

The trouble is, there is still a multitude of these non-quantum histories. They are described by predictable, non-quantum laws, of which the most important – because it describes how the large-scale Universe evolves in time – is Einstein's theory of gravity.

Einstein's theory describes all kinds of possible universes. There are ones that are clumpy or smooth, ones that re-collapse after a short time or expand for ever, ones that expand at breakneck speed or at a snail's pace, and so on. Our Universe is one universe among this throng, but there appears to be nothing distinctive about it. Nothing to make it stand out from the crowd. Nothing to tell us why we have ended up in our particular Universe rather than any other. What is needed is another plausible reason to thin out the forest of possible cosmic histories. And such a reason has been proposed by physicists Stephen Hawking and James Hartle.

The picture of multiple histories of the Universe all bundled together turns out to be only half the picture. Such histories are properly determined – anchored in reality, if you like – only if the conditions at the beginning of time are pinned down. Unfortunately, physicists are as ignorant of the 'initial quantum state' of the Universe as they are of the theory of quantum cosmology itself. At least, they were. In the 1980s, Hawking, of the University of Cambridge, and Hartle, of the University of California at Santa Barbara, noted something interesting about Einstein's theory of gravity: it can be reformulated in such a way that instead of having three dimensions of space and one of time, it has three dimensions of space and one of 'imaginary time'.

Imaginary time is a mathematical concept which it is not necessary to understand. The key thing is that it behaves just like space. Using this insight, Hawking and Hartle were able to show that in the initial quantum state, the multiple histories of the Universe, instead of existing in space and time, could have existed in space alone. This allowed them to sidestep neatly the sticky question of what happened before the Big Bang. After all, if the Big Bang happened in space alone – outside of time – asking what happened *before* the beginning is like asking what it is like north of the North Pole. There is nothing north of the North Pole. It is a question with no meaning.

Remarkably, this means that the initial condition of the Universe could have been that *there was no initial condition.* Hawking and Hartle have dubbed this the 'no boundary condition'. It provides a way of further thinning out possible cosmic histories. When possible histories are reformulated in terms of space alone and the no-boundary condition applied, some of those histories turn out to have an extremely small chance of ever occurring. They can therefore safely be discarded. And this is what Hawking and Hartle, together with their colleague, Thomas Hertog, of the University of Paris, did in late 2007. They then looked at the cosmic histories that survived the cull. To their surprise, all of the survivors shared a striking feature: at the outset each underwent a period of super-fast expansion.

This was a very significant discovery. A burst of super-fast expansion is the preferred way that cosmologists fix a serious problem with the Big Bang model. In a nutshell, the basic Big Bang model does not work. It predicts something we do not see when we look out at the Universe. According to the model, the Universe began in a super-dense, super-hot state about 13.7 billion years ago and has been expanding and cooling ever since, with the galaxies and stars congealing out of the debris. However, this simple picture makes a prediction which is dramatically at odds with what we see. It concerns the 'cosmic background radiation', the leftover heat of the Big Bang fireball.[3] Greatly cooled by the expansion of the Universe over the past billions of years, the afterglow of the Big Bang permeates all of space and comes to us directly from an epoch about 380,000 years after the beginning of the Universe. And herein lies the problem.

If we imagine the expansion of the Universe running backwards to this epoch like a movie in reverse, we find that our currently observable universe was then about 20 million light years across. This means it was impossible for any unevenness in temperature that developed to be ironed out. After all, to even out the temperature, heat would have to flow from warm regions to colder regions. But the Universe was too big. Even at the cosmic speed limit – the speed of light – heat could have travelled no more than 380,000 light years since the Big Bang, and this was only a tiny fraction of the way across the Universe at that time.

A prediction of the basic Big Bang model is therefore that the temperature of the cosmic background radiation cannot possibly be the same in different directions in the sky. The trouble is, it is. To within a tiny fraction of a degree, wherever a telescope is pointed in the sky, it is measured to be 2.7 degrees above absolute zero.

What this contradiction is telling us is that the basic Big

Bang model must be an incomplete description of the Universe. Something else is needed. Something new must be bolted on.

One possibility is that there was a long pre-Big Bang era. If this was the case, then the cosmic temperature would have evened itself out automatically. It would be like running cold water into a hot bath. If you wait long enough, the temperature becomes uniform. Another possibility, championed by the physicist João Magueijo, is that the speed of light was much greater in the Big Bang than it is today. This would have enabled heat to have travelled from hot regions of the fireball to colder regions far faster than expected, again evening out the temperature. But there is another possibility, and this is the one that has been embraced by most physicists: that the Universe underwent a phenomenal burst of expansion in its first split second of existence.

'Inflation' was proposed by the Russian physicist Alexei Starobinsky in 1979, and independently by the American physicist Alan Guth in 1981. It has been likened to the detonation of an H-bomb compared with the stick of dynamite of the Big Bang expansion that followed inflation when it ran out of steam. If inflation did occur, then 380,000 years after the Big Bang the Universe would have been far smaller than we deduce from running the movie of its expansion backwards, small enough for heat to travel back and forth easily, ironing out the cosmic temperature.

The trouble with the inflationary idea is that it is untestable practically. The conditions of temperature and density that existed in the first split second of the Universe are so extreme that we could never reproduce them in the laboratory to check what happens. Not only this but, in the three decades since inflation was first proposed, nobody has come up with a compelling explanation of why it happened. The mechanism that underpins it is a mystery. Inflation is simply tagged onto the basic Big Bang model in an unsatisfactory ad hoc manner.

This is why the deduction made by Hawking, Hartle and Hertog is so significant. Simply by taking the mundane observation that we live in a largely quantum Universe and applying their no-boundary condition, they have shown that the most likely possible histories of the Universe involve super-fast expansion. Inflation simply must have happened. It is unavoidable.

But why should the conditions imposed by Hawking's team on possible universes pick out only those that undergo inflation? The answer is that inflation provides the most likely route for a universe to go from quantum to non-quantum. Quantum tends to be synonymous with small. Non-quantum tends to be big. The quickest way to get from the small to the big is with a burst of super-fast expansion. Unfortunately, there is a problem. Although the team's analysis shows that inflation was unavoidable, it also shows that it was very short-lived. Too short-lived. The Universe would have doubled its size only a few times over, whereas observations of our Universe reveal that inflation actually doubled its size more than 60 times over, causing it to mushroom by a truly mind-blowing factor.

The inflation deduced by Hawking's team is short-lived because inflation involves the Universe starting out in an unstable state. Nobody knows exactly what matter 'field' is responsible for the instability, although cosmologists commonly talk of a hitherto undetected 'inflaton' field whose repulsive gravity inflates the Universe. The point, however, is that anything in an unstable state tends to want to return to stability, and this is more likely to happen quickly than after a long time. Think of a pencil balanced on its tip. Clearly, this is a highly unstable situation. Buffeted by a draught and vibrations, the pencil is far more likely to keel over after a fraction of a second rather than after a day and a half. Similarly, the instability of inflation is far more likely to end after a short time than after a long one.

Fortunately, Hawking's team has recently realised that there is a way to rescue their idea. It turns on the fact that we see only a tiny part of the Universe that inflated. This is because the Universe is just 13.7 billion years old, so we can see only those objects whose light has taken less than 13.7 billion years to reach us. Light from the rest of the Universe has not arrived yet on Earth. It is beyond the horizon, an imaginary boundary which surrounds us like the surface of a bubble. The bubble is, of course, the observable universe, as mentioned before. But there are other bubbles out beyond the horizon of the observable universe. Somebody else's observable universes. Astronomers call them Hubble spheres.

There are a large number of Hubble spheres. In fact, the number is roughly equal to $e^{[\text{number of doublings during inflation}]}$.[4] And although we find ourselves in this particular Hubble sphere, we could equally well have ended up in the next Hubble sphere. Or the next.

But here is the crucial thing. The more doublings the Universe has undergone, the more possible places we could find ourselves within it. So although universes which undergo more doublings are less likely – because the instability of inflation is more likely to end after a short while than a long while – there are more places in such universes where we could find ourselves. And this effect, it transpires, wins out. So contrary to what Hawking's team originally thought, it is overwhelmingly likely that we will find ourselves in a universe which underwent a long period of inflation rather than a short one. At long last, we appear to have an explanation for why we find ourselves in a universe which inflated.

Perhaps it is pushing it a little to say that the non-quantumness of the everyday world is telling us that the Universe must have undergone a period of super-fast expansion in the past. But according to Hartle and his colleagues, this is exactly what the non-quantumness of the everyday world *combined with* the no-boundary idea is telling us. The evidence that inflation occurred is all around you: in the fact that the world is predictable and that, when you walk past a tree, you walk past on one side and not on both sides simultaneously. But although we have come to the end of this chain of reasoning, there is still one loose end. A rather important loose end.

How exactly do histories that are non-quantum at late times – the only ones we considered – become non-quantum? It is all very well to say that bigness is associated with a universe being non-quantum. But how exactly does a universe make the transition from being quantum to non-quantum? This is one of the most fundamental questions in science. After all, we live in a universe orchestrated by quantum theory and yet nowhere – at least in the everyday world – is quantum behaviour obviously apparent.

The key to resolving the paradox is a process called 'decoherence'. For quantum behaviour to manifest itself, it is fundamental that the quantum probability waves representing the possibilities open to an object mingle with each other, or 'interfere'. This is because interference is at the very root of quantum weirdness. If quantum waves do not interfere with each other, there is no quantum weirdness. Waves that can interfere with each other are said to be 'coherent', which is why the process by which they lose this ability – and lose their quantumness – is called decoherence.

In the early Universe, decoherence occurred in the following way. Quantum uncertainty – Heisenberg again – caused the properties of things to fluctuate wildly.[5] Take space–time itself. Close up it resolved itself into violent contortions like a choppy, storm-tossed sea. The enormous expansion of inflation stretched, or magnified, this choppiness. It turned space–time into a landscape with gently sloping hills and valleys. The valleys were places that particles of matter gradually fell into, and the hills were places they avoided. By this process, the structures of today's Universe, such as giant clusters of galaxies, began to grow. And it was this process – the clustering of matter – that triggered the transformation from a quantum to a non-quantum universe.

Focus for a moment on a valley at a particular location. Such a valley is a quantum entity – magnified by inflation but a quantum entity nonetheless. And like all quantum entities it has many possibilities open to it. For instance, in one possible history of the Universe, the valley is present. And in another, it is not. But, crucially, both possibilities can exist simultaneously, rather like Schrödinger's eponymous cat being dead and alive at the same time. Which means the quantum waves representing the possibilities can interfere with each other.

Now imagine a particle of matter that falls into the valley. The same particle exists in the alternative history in which there is no valley, so it has nothing to fall into and stays put. Therefore, the particle is in slightly different locations in the two possible universes. The quantum wave representing one case does not quite overlap with the quantum wave representing the second case. Say, for the sake of argument, there is only a 50 per cent overlap.

Now consider a second particle joining the first in the valley. Again, this means the particle is in different locations in two possible universes. Say the quantum waves for the two cases overlap by 50 per cent again. This means that the overlap between the quantum waves representing the two particles in the valley and not in the valley is $\frac{1}{2} \times \frac{1}{2} = \frac{1}{4}$. With a third particle falling into the valley, the overlap is $\frac{1}{2} \times \frac{1}{2} \times \frac{1}{2} = \frac{1}{8}$. See where this is going? With each successive particle that falls into the valley, the quantum wave representing the history where the valley is there overlaps less with the wave representing the history where the valley is not there. Once we get to trillions upon trillions of particles, there is essentially no overlap. There can be no interference. And so the Universe has become non-quantum.

It was the growth of clumps of matter, leading to galaxies and you and me, that therefore caused the transition from quantum to non-quantum. But this still leaves one big mystery. Merely talking of fluctuations in space–time assumes the *pre-existence* of space–time, a smooth entity like a glassy sea that could be ruffled by Heisenberg uncertainty. Where did this smooth non-quantum space–time come from? That is indeed the big question. Clearly, it must have arisen from something totally quantum, something so turbulent and chaotic that space and time were meaningless concepts. This *thing* was the quantum precursor of space–time. It must have undergone a transformation to the non-quantum space–time we see around us today. How this happened nobody yet knows. There are no shortcuts to understanding the origin of space–time. It will require a true quantum theory of the Universe.

9

The Humpty Dumpty Tendency

*How the fact that teacups break but never unbreak is telling us
that the Universe must have expanded from a big bang*

> 'Humpty Dumpty sat on a wall.
> Humpty Dumpty had a great fall.
> All the king's horses and all the king's men
> Couldn't put Humpty together again.'
>
> Nursery rhyme (unknown origin)

> 'Lettin' the cat outta the bag is a whole lot easier 'n puttin' it back in.'
>
> Will Rogers

*A teacup slips from your grasp. It hits the floor and shatters into
a dozen pieces. Such an everyday accident may seem of little
consequence, but it is telling us something profound about our
Universe. It is telling us that the Universe must have begun in a
highly unusual, ordered state, one which is possible if it
expanded from a Big Bang explosion.*

How can the birth of the Universe in the Big Bang have anything to do with something so prosaic as a teacup shattering?
The answer is subtle. And it explains not only why teacups
break rather than unbreak but why tea left in a cup grows
cold rather than hot and even why you grow older rather
than younger with every passing year.

Think of that disintegrating teacup. It illustrates perfectly
a paradox at the heart of the world, one which for a long
time baffled physicists. The paradox arises because of a striking feature of the fundamental laws of physics which orches-

trate the Universe and decree what can and cannot happen. If those laws permit a particular process to happen, they always permit its opposite too.

Take Newton's laws of motion, which describe the movement of bodies under the influence of forces. A manoeuvre often used by the American space agency NASA to boost a space probe so it can reach the outer Solar System is a so-called sling-shot. A space probe, moving under the influence of the Earth's gravitational force, approaches the Earth, swings around the planet and boomerangs back out into the depths of space. Imagine you were shown a movie of this fly-by and a movie of the fly-by running backwards. In the latter case, the space probe approaches the Earth from the direction in which it formerly receded and recedes along the direction it approached. The question is: could you tell simply by looking which depicted the manoeuvre actually used by NASA?

The answer is no. If a certain trajectory for a spacecraft is permitted by Newton's laws of motion, so too is its opposite. The laws are 'time-symmetric', which means you can never tell whether the movie is running forwards, as normal, or backwards. Both situations are equally permissible.[1]

What has this got to do with a shattering teacup? Well, each fragment flying through the air is as much under the influence of Newton's laws of motion as a space probe swinging around the Earth. Say you were shown a movie of a single shard tumbling through the air and the same movie running backwards. Assuming the camera had focused only on the shard and nothing else tell-tale in the surroundings, could you pick the movie which depicted reality and the one which showed reality running backwards? As with the space probe, the answer is no.

But now zoom out from the fragment to the teacup

shattering on the floor. If you were to see a movie of this event and a movie of the event running backwards, could you tell which was reality and which was reality in reverse? The answer is, of course, yes. In the real world, teacups shatter. They never unshatter. But how can a teacup whose component fragments are governed by fundamental laws that are time-reversible behave in a way that is so emphatically not time-reversible? This is the paradox that for so long baffled the world's best physicists.[2]

And it is not just teacups that behave like this. If you were to watch a movie in reverse of any event in the everyday world, you would know instantly which was reality and which was not. People run after buses, they do not run away from buses; dogs chase cats, they do not un-chase cats; people grow old, they do not grow young. So how is it that we can live in a world orchestrated by fundamental laws that permit things to happen equally well backwards or forwards in time and yet be surrounded by events which happen only forwards? The man who provided the language to discuss the paradox in a precise way was the Austrian physicist Ludwig Boltzmann.

It would be nice to relate a quirky or amusing anecdote about Boltzmann. Unfortunately, he had a pretty tragic life. It was a shame because, by all accounts, the short, stout, bespectacled Boltzmann, known to his fiancée as 'my sweet fat darling', was kindly and well liked. He could never refuse a request for a favour and cared so much for his students, especially those of poorer means, that towards the end of his life no student was ever permitted to fail one of his examinations. But Boltzmann was dealt a number of personal blows. His father died when he was 15, a time in the life of a child that is well known to be the worst to lose a parent, and his first son, Ludwig, died of a burst appendix. Boltzmann blamed himself for not recognising early enough the seriousness of his 11-year-old's condition, though it is hard to imagine that, in an age before antibiotics, anything would have made a difference. These events probably played a role in triggering, at the age of 44, Boltzmann's manic depression, although they probably just exacerbated an existing, though not so obvious, condition. For the rest of his life Boltzmann oscillated between periods of euphoria when he felt himself the master of the world – or at least of the world of physics – and often worked feverishly until 5 a.m., and episodes of terrible black despondency when all his triumphs seemed but worthless dust and ashes. It was in 1906, during one of these episodes of depression, which coincided with a family holiday in Duino, a village on the Adriatic coast near Trieste, that his 15-year-old daughter Elsa – the sunshine of his life – returned from swimming to find him hanging in his room. Boltzmann was 62.

Undoubtedly, a contributory factor in Boltzmann's suicide was the hostility directed at him by an army of scientific zealots who considered that atoms, which nobody had ever seen or touched and on which Boltzmann's ideas were founded, were a dangerous idea that must be ruthlessly rooted out of science, lest they bring the whole edifice tumbling down. The irony is that only the previous year Einstein had shown how the mysterious jittery motion of pollen grains suspended in water could be explained if they were coming under constant machine-gun bombardment from atoms in the water, and, in 1908, Jean Baptiste Perrin would even use this Brownian motion and Einstein's theoretical framework to measure the size of atoms. Atoms were no fiction, as Boltzmann's enemies claimed. They were real. The final tragedy of Boltzmann's life was that he did not experience the glorious victory of his ideas, nor was he around to see his elevation to the pantheon of the greatest physicists who ever lived.

It was one of Boltzmann's greatest triumphs to show how time-symmetric laws of physics could lead to an everyday world with a very definite direction of time. The key, he realised, was probability.

Think of the teacup again. Specifically, think of how many ways it could shatter. It could shatter, for instance, into ten fragments, or 19, or 48. It could shatter into a small number of big fragments and a large number of small fragments. It could shatter exclusively into small fragments. And so on. It does not take much imagination to realise that the number of possible ways a teacup can break is monumentally huge. Now, among all these possibilities, is the possibility that the teacup does not shatter at all but remains intact. But there is only one way this can happen. Consequently, if all outcomes for the teacup are equally likely – and this was Boltzmann's reasonable assumption – then it is overwhelmingly probable that the teacup will shatter.

Now imagine the event in which a teacup unbreaks. Say you drop a collection of cup fragments, they hit the floor and come together miraculously to make an unbroken teacup. In how many ways could this happen? Well, as pointed out before, there is only one way the cup can be intact. Contrast this with the astronomically huge number of ways that the cup can remain broken: it can stay in the same number of pieces, or all of its pieces can shatter into even smaller pieces, or some of its pieces can stay intact and others crumble to dust, and so on. The point is that there are hugely more ways the cup can remain broken than reassemble into an unbroken cup. Once again, if all possibilities are equally likely, it is overwhelmingly probable that we will not see the shards of the teacup leap back together again to make a pristine vessel. It is not totally impossible; however, it is an event so mind-bogglingly improbable that you would have to keep dropping teacup shards over and over, for far longer than the current age of the Universe, before you were lucky enough to witness such an extraordinary event.

Such improbable events turn out to have huge philosophical implications if the Universe is infinite in extent, either in space or in time. After all, in such a universe anything, no matter how mad and unlikely, is certain to happen. The trouble is, we appear to live in such a universe. In 1998, physicists and astronomers in the US and Australia discovered that the expansion of the Universe is speeding up, driven by 'dark energy', invisible stuff with the repulsive gravity that fills all of space. It appears that the dark energy may cause the Universe to grow without limit. Nobody knows what the dark energy is. However, like all things quantum, it will inevitably undergo quantum fluctuations, conjuring particles of matter out of nothing. Such quantum fluctuations could create a man in a space suit floating in space. Or a computer. Or a brain with a single giant eye. These possibilities are, of course, mind-bogglingly unlikely. However – and this is the crucial point – in a universe with an infinite amount of space and an infinite amount of time, they are certain to happen; in fact, they are certain to happen an infinite number of times over. The problem is that eventually such 'Boltzmann brains', as they are known, will outnumber ordinary observers like you and me who have evolved over billions of years by the hand of natural selection. This is a problem because our models of the Universe – the Big Bang models – are founded on the idea that we are typical observers and that what we see as we look outwards at the Universe is typical of what all cosmic observers see. If most observers are Boltzmann brains, staring out at unending tracts of utterly empty space, then they are the typical observers. The foundation stone of our cosmology would crumble, and with it everything we thought we understood about the Universe.

But back to teacups. What characterises the state of the intact teacup is 'order', while the state of the shattered teacup is characterised by 'disorder'. Physicists have a technical name for disorder: 'entropy'.[3]

Boltzmann's key insight was that when a body has a large number of subcomponents, the disordered possibilities open to it vastly outnumber the ordered possibilities. If all possibilities are equally likely, there is therefore an overwhelming tendency for the body to become more disordered with time – for it to increase its entropy. Since we associate the direction in which order becomes disorder with the direction of time – a teenager's bedroom tends to get more untidy with each passing day, not more tidy – this neatly explains why in the everyday world there is a direction, or 'arrow', of time. It shows how, despite the fact the underlying laws of physics permit the subcomponents of a body to do things equally well forwards or backwards in time, the body itself always behaves like a forward-running movie.

Everything around us is made of subcomponents – atoms. You are made of about 1,000 billion billion billion, and it would take 10 million, laid end to end, to span the full stop at the end of this sentence. Since the number of atomic subcomponents is so enormous, it is not simply teacups shattering into a thousand pieces that are characterised by a change from order to disorder. So too are all everyday processes.

Changes from order to disorder in your surroundings may not appear obvious. Someone claps their hands. How does this increase disorder? Or, to take a more esoteric example, someone fires a bullet into a steel wall. How does this boost entropy? The answer is that such processes add to the disorder of the Universe in a subtle way – by producing 'heat'.

Heat is actually random microscopic motion. For instance, it is the jiggling of atoms about their average locations in a 'solid'. It is the frenzied motion of atoms like a swarm of angry bees in a 'gas' such as air (such atoms fly about according to Newton's laws of motion just as surely as a space probe flying through space). It is the random machine-gun sputter of particles of light, or photons, emitted by the atoms of a hot body.[4] Heat is the very epitome of disorder.

In the case of a person clapping their hands, the concussion of flesh jiggles the air molecules in the immediate vicinity, which jiggle their neighbours, exporting disorder into the surrounding air. It heats up a little. In the case of the bullet burying itself in the steel wall, friction jiggles the atoms in not only the wall but the bullet itself. They heat up a lot.

When any form of energy is turned into heat energy, it is as unlikely for that heat energy to be turned back into the original form as it is for a teacup to unbreak. And for the same reason. In the case of the person clapping their hands, it would require the zillions of jiggling air atoms in the thin layer between the person's hands suddenly to find themselves jiggling outwards in perfect unison so as to push the hands apart again. Though not totally impossible, this is a fantastically unlikely thing to happen. And in the case of the bullet embedded in the steel, the jiggling atoms of the bullet would all have to find themselves jiggling in the same direction – away from the steel wall – so that the bullet leapt back out towards the gun. How likely is that? The answer, of course, is overwhelmingly unlikely.

All this leads to a fundamental asymmetry in the law of conservation of energy when it involves heat energy. Although it is possible to convert 100 per cent of any form of energy into heat energy, it is not possible to turn 100 per cent of any heat energy into another type of energy. The reason is that heat represents energy dispersed among many, many things – for instance, among the many molecules in a hot gas. To convert this energy into useful 'work' involves concentrating energy into a few things – for instance, the motion of a single bullet. The reason heat cannot be changed into work with 100 per cent efficiency is simply that it is extremely unlikely for the energy in a hot substance, which is spread among many, many states, to pass into a few states, such as the motion of a bullet.

It is not at all obvious in the case of the bullet and the steel wall – or in the case of the clapping hands – that disorder has increased. In fact, it would appear that in both instances the opposite has happened. After all, two hands glued together seems a more ordered state than two hands held apart. A bullet and a steel wall welded into one entity is more ordered than a bullet and a wall as two entities. What is not obvious to us – because it is invisible to the naked eye – is that these local increases in order have been more than paid for by an increase in global disorder – by the export to the environment of microscopic disorder, or heat.

The fact that order can increase locally at the expense of disorder being boosted elsewhere hints at how humans and all other living things on Earth have bucked the trend of ever-increasing entropy. A baby feeds on its mother's milk and the energy this provides enables a proliferation of neuronal connections in its brain, creating the most ordered entity in the known Universe. But in metabolising that milk, heat is produced – we all know we get warm when we eat – and the environment pays. Overall, the disorder of the Universe increases.

The fact that disorder can never decrease is known as the 'second law of thermodynamics'. Although it is an inviolable law – any law in physics that is found to contradict it is instantly discarded – it is actually on a different par to other fundamental laws of physics. Whereas they determine what happens with 100 per cent certainty, the second law is statistical. It ordains only what is overwhelmingly likely to happen, rather than what is certain to occur.

One of the consequences of the transformation of order into disorder in the Universe – of the remorseless increase of entropy – is a running down of things. A constant degeneration. A degradation. Ultimately, it explains why we grow older – how the function of our cells becomes ever more compromised by disorder. It is the hand of entropy at work as surely as in the shattering of a teacup.

But if all processes in the Universe increase its net disorder, then a logical conclusion that can be drawn is that one day the Universe will reach a state of maximum disorder. In such a universe, it will be impossible for anything at all to happen since, as pointed out, no everyday process can happen without an increase in the Universe's disorder. Even worse, the past and the future will lose all meaning since we identify the direction from past to future – the arrow of time – with the direction in which things become increasingly disordered.

That there exists a hypothetical state of maximum disorder for the Universe was first recognised by physicists in the nineteenth century. They christened it 'heat death', and it is the nightmare scenario when all activity grinds to a halt. The Universe, in the words of T. S. Eliot, ends 'not with a bang but a whimper'.

The way to understand precisely why nothing can happen in a state of heat death is to realise that not all disorder is equal. There is usable disorder and unusable disorder. It all depends on a particular property of heat: its temperature, or degree of hotness.

Most people recognise the difference between heat and temperature. A lighted match has a high temperature but contains little heat – try using it to heat a saucepan of water – while a household radiator contains a lot of heat but has a low temperature – if you touch it you will not burn yourself. And it turns out that if you have two sources of heat at different temperatures, it is possible to do useful 'work' – for instance, to drive a piston in a steam engine. It is exactly like having a difference in water level, such as in two ponds at different heights. If water flows from the higher pond to the lower one, it can turn a water wheel. But the inevitable result of this is that all the water ends up at the same level in the lower pond and, with no difference in water level, the water wheel can no longer be turned.

The same thing happens in a steam engine. Steam at high temperature drives a piston and then is condensed, or turned to water, and discharged into the atmosphere at a lower temperature. Since the discharged heat is at the same temperature as the atmosphere, it is spent and can do no more work. And this is what physicists of the nineteenth century realised would happen to the Universe when it reached heat death.

The key temperature difference that drives everything in the Universe is the one between the stars, which are hot, and empty space, which is cold. All processes on Earth ultimately derive their energy from sunlight, which is radiated into space because there is a temperature difference of about 6,000 degrees between the solar surface and the surrounding space. For instance, sunlight drives the circulation of the ocean and atmosphere. It drives photosynthesis in plants. Ancient microorganisms and trees soaked it up and, after being squeezed beneath the ground for millions of years, became fossil fuels such as oil and coal.

But as the stars pump heat into space, they inevitably get cooler and space gets warmer.[5] Eventually, the temperature difference between them will be ironed out. The Universe will be filled not only with disorder but unusable disorder. Disorder at the same temperature. This is the dreaded state of heat death.

So how close is the Universe to such a state? Well, space is criss-crossed by photons pumped out randomly by stars. It turns out that their contribution to the disorder of the Universe far outweighs that of all the jiggling particles of matter put together. And even the photons from stars account for a mere 0.1 per cent of the photons currently abroad in the Universe. An enormous 99.9 per cent are tied up in the left-over heat of the Big Bang fireball – the afterglow of creation. In fact, there are 10 billion of these photons for every particle of matter. Hold up your hand. Every second a million billion photons from the Big Bang are bouncing off your skin.

The leftover heat of the Big Bang is telling us that the Universe today is essentially in a state of heat death. Orbiting a star, we exist in a rare cosmic location where activity can still go on. Shockingly, it seems that most of the Universe succumbed to heat death well within the first second of its existence. It happened when an orgy of particle–antiparticle annihilations, each of which created a pair of photons, destroyed most of the Universe's matter and antimatter. Because some as-yet-not-understood asymmetry in the laws of physics had ensured that for every 10 billion particles of antimatter there were 10 billion and 1 particles of matter, when the dust finally settled there were about 10 billion photons for every surviving particle of matter. It is these photons we see today as the afterglow of the Big Bang – cosmic background radiation.[6]

The fact that the Universe is currently so close to heat death means that there is actually very little scope for disorder to increase in our Universe, so it is not possible to explain why time flows the way it does – why teacups shatter and people grow old – by the Universe being far away from a state of heat death. What, then, is the explanation for the arrow of time? Well, it is a statement of the blindingly obvious, but a teacup can become more disordered in the future only if it was more ordered in the past. Consequently, the fact that time flows the way it does is telling us that in the Big Bang the Universe must have been in a highly ordered, highly improbable, special state.[7]

Physicists have an abhorrence for accepting that there is anything special or unlikely about our Universe. It smacks of religion. It goes back to Nicolaus Copernicus's discovery that there is nothing special about our position in the Universe – the Earth orbits the Sun along with the other planets and is not the centre of things – and Charles Darwin's discovery that there is nothing special about our place in the natural world – human beings are just one among myriad other animal species on our planet. However, to make sense of the arrow of time we experience, physicists have been forced to accept that the Universe must have started in an unlikely, highly ordered state. Most harbour the hope that such a state will one day be shown to be an inevitable consequence of a 'theory of everything' that explains all of creation in one neat set of equations.

However, one physicist thinks that it is not necessary to look to the super-physics of an elusive theory of everything to explain how the Universe started out in an unlikely, ordered state. According to Lawrence Schulman, of Clarkson University in New York, there is a relatively mundane explanation. Regardless of how the Universe started out, he believes, it naturally made a transition to an unlikely, low-entropy state when it was about 380,000 years old. The key was gravity, which, for the first time, gained control of the Universe.

The era 380,000 years after the moment of creation was of crucial importance in the history of the Universe. At this time, the expanding fireball of the Big Bang had cooled to about 4,000 degrees, a sufficiently low temperature for atomic nuclei and electrons to combine to make the first atoms.[8] Before this time, photons – of which there were about 10 billion for every particle of matter – had a powerfully disruptive effect on matter. They ricocheted, or 'scattered', off free electrons, blasting them apart and so preventing gravity from gathering together any clumps of matter. After that time, electrons were bound up inside atoms and so relatively shielded from the disruptive effect of photons. For the first time, gravity could begin to cause matter to clump. It was the start of a long process that would lead ultimately to galaxies, stars, planets and you and me.[9]

It was the 'switch-on' of gravity at this 'epoch of last scattering', 380,000 years after the birth of time, that Schulman believes explains why the arrow of time points in the direction we know and love. Before this epoch, the glowing matter of the Big Bang fireball was smeared uniformly throughout space. In fact, present-day observations of the cosmic background radiation – the expansion-cooled photons of the Big Bang fireball – tell us that matter at that time was smooth to a level of about one part in 100,000. A state as smooth as this turns out to be the most likely high-entropy state when there are no long-range forces. Think of a gas in a box. The atoms of the gas, flying about freely, spread themselves evenly throughout the box.

However, once a long-range force – gravity – switched on in the Universe, things changed radically. The most likely state for matter in the presence of gravity is not a smooth state but a clumpy one. Just look around at today's Universe, filled as it is with galaxies and stars and planets. When gravity switched on at the epoch of last scattering, therefore, the smoothed-out state of the Universe suddenly switched from being highly likely to highly improbable.

It is subtle. The distribution of matter was exactly the same before and after the transition. However, what had been a high-entropy, typical state when gravity was playing no part in events suddenly became a low-entropy, special state when gravity entered the game.

Schulman freely admits that his argument says nothing about the direction of time before the Universe was 380,000 years old. It is possible that there was no arrow of time. Or maybe there was an arrow of time set by some earlier physical processes. Figuring out what the arrow of time was doing at such impossibly remote times may indeed require a theory of everything.

However, by showing how the Universe could have made a transition to a special, low-entropy state early on, Schulman appears to have finally explained why we experience the direction of time we do.[10] And it is all because of an event caused by the expansion-driven cooling of the Universe, for it was this that led to the formation of atoms and the 'switching on' of gravity. It was the latter event that put the Universe in the special, low-entropy state so essential for explaining the observed arrow of time. The fact you grow old, not young, that your coffee gets cold, not hot, that eggs break rather than unbreak is therefore telling you that the Universe must have expanded from the Big Bang. There can be few better examples of how the cosmic is connected to the everyday.[11]

10

Random Reality

How the fact that lots of information is needed to describe the world tells us chance played a key role in creating everyday reality

'Information is a revolutionary new kind of concept and the recognition of this fact is one of the milestones of this age.'
Gregory Chaitin (*The Unknowable*)

'We profess ourselves to be the slaves of chance.'
William Shakespeare (*The Winter's Tale*)

The world is complex. Rain clouds scud across the sky. A tree sways gently in the breeze. A woman in a red coat walks her cream poodle down the street and stops at a pedestrian crossing. Describing such a scene precisely requires a vast amount of information. It is necessary, for instance, to specify the location, shape and composition of every cloud, the location and shape of every branch and leaf on the tree, and so on. The reason it takes so much information is that a lot of things have to be specified in order to ensure that the scene is uniquely distinguishable from myriad other possibilities. This is because there are an awful lot of ways the scene could be different, an awful lot of alternative ways its 'stuff' could be arranged. A cloud could be in another place, a lamp post could substitute for the tree, a man walking his pet ferret could replace the woman and the dog. In fact, to be sure the scene cannot be mistaken for any other possible scene, it is necessary to specify the location and properties of every single atom in the scene. Every subatomic particle even.

The observation that a vast amount of information is needed to describe the Universe may seem trite and of little consequence. But actually it is telling us something profound about our Universe. According to physicist Stephen Hsu, of the University of Oregon in Eugene, it is telling us that the world around us is the way we find it and not some other way because of pure chance. It is telling us that the complexity of the world is the outcome of a long series of rolls of the dice extending all the way back to the beginning of time. Einstein famously declared that 'God does not play dice with the universe.' But, says Hsu, 'Not only does God play dice with the universe but, if he did not, there would be no universe – at least, not one of the richness and complexity for life to have arisen.'[1]

How is it possible to come to such a startling conclusion merely from the fact that the Universe is complex and so requires a lot of information to describe it? Well, strictly speaking, it isn't. Such a conclusion is possible only by comparing how much information there is in today's Universe with how much there was when the Universe was born.[2]

According to the standard picture of cosmology, the Universe, with all its billions upon billions of galaxies and stars, inflated from a tiny piece of 'vacuum' far smaller than an atom. Estimating the information content of this pre-inflation patch is technical, but Hsu uses the following simplified argument. Before inflation, when the Universe was about 10^{-44} seconds old, the four fundamental forces of nature are believed to have been 'unified' into a single 'super-force'. The Universe at this epoch was at the 'Planck temperature' – about 10^{32} degrees – and was no larger than the 'Planck length' – about 10^{-35} metres. Lacking as they do a quantum theory of the force of gravity, physicists can know nothing about this time. However, by the time the Universe had expanded to ten times the Planck length and its temperature had fallen to a tenth of the Planck temperature, the effects of quantum gravity were already minimal. It is therefore possible to say something about this time. The Universe can be imagined as made up of 1,000 cubes, each with sides of the Planck length. The reason for thinking this is that the Planck length is the minimum possible length, akin to the dot size on a newspaper photograph. As already pointed out, the information content of something is related to the number of distinguishable ways that the subcomponents of the thing can be arranged. So the key question is: how many distinguishable states were there in this 1,000-cube universe?

Each of the small cubes could be either filled with energy or be empty, just as a location on a photograph can be filled with black ink or be empty. A universe made of 1,000 filled or unfilled cubes is not easy to visualise, so think of a one-cube universe. The cube could be either filled or empty, making two distinguishable states. In a two-cube universe, the cubes could be empty–empty, empty–full, full–empty or full–full, making four possible arrangements, which is equivalent to 2^2. In a three-cube universe, there are 2^3 arrangements. In a four-cube universe, 2^4. See the pattern? It is therefore possible to say that in the 1,000-cube universe that existed prior to inflation, the number of distinguishable ways the vacuum could have been arranged was $2^{1,000}$.

$2^{1,000}$ – 2 multiplied by itself 1,000 times – is approximately a billion. It may seem like a lot of possible arrangements for the pre-inflationary patch of vacuum, but actually it is ridiculously small. Think of a computer disk. When people say a disk can store N bits of data, they mean the disk can store 2^N different strings of os and 1s, or binary digits (bits).[3]

Listing all the possible ways that the stuff of the pre-infla-tionary patch could have been arranged therefore requires a mere 1,000 bits of storage. In other words, the precise state of the early Universe would fill only a kilobit of disk space. A byte – 8 bits – is usually used to store a character such as an 'A' or a '5', so specifying a whole universe requires less than 200 bytes. Imagine being given a piece of paper with 200 characters, or about 30 words, scrawled across it. Incredibly, this is sufficient to specify the state of an entire universe. If this is not mind-blowing enough, think of it another way. In *Song of Myself*, Walt Whitman wrote: 'And I say to any man or woman, Let your soul stand cool and composed before a million universes.' Well, today it is easy to 'stand cool and composed before a million universes'. Just buy a 1 gigabit (Gb) key-ring flash memory. Believe it or not, you could store the information for a million universes on it.

The idea that the $2^{1,000}$ possible arrangements of the pre-inflation vacuum can be stored in 1,000 bits highlights the definition of information. If the number of possible arrangements of something is 2^N, then the information con-tent is defined as N.[4]

So much for the piddling information content of the pre-inflation Universe – but how much information is there in today's Universe? In calculating the figure, the key thing is to recognise where most of that information resides. Even if we possessed a super-telescope that could look out at the Universe and record the precise state of every atom in every star in every galaxy, there would still be an overwhelmingly large number of ways the Universe could be different from our specification. The reason is that space is permeated by a vast number of photons from the Big Bang – the leftover heat from the primordial fireball. The photons of the Big

Bang outnumber microscopic particles of matter such as electrons by a factor of about 10 billion and even photons of starlight by a factor of 1,000. Their precise state is therefore the biggest unknown in the Universe.

The Big Bang photons are indistinguishable, so swapping them does not result in a new arrangement. However, a photon has two distinct states open to it because it can be 'polarised' in two different ways. Think of a photon flying through space as corkscrewing either clockwise or anticlockwise about its direction of motion. If we imagine each photon as traversing its own cube of space, then the situation is pretty similar to the pre-inflation vacuum, with each cube containing either a clockwise photon or an anticlockwise one rather than being either filled with energy or empty. The same result, therefore, holds. The number of distinguishable states of N Big Bang photons is 2^N,[5] so the information required to describe the states of all the Big Bang photons in the Universe is simply N, the total number of those photons. At any moment every cubic centimetre of space is being traversed by about 300 relic photons – that's how ubiquitous the photons of the Big Bang are. The Universe is about 84 billion light years across, which means it has a volume of about 5×10^{86} cubic centimetres.[6] Consequently, the amount of information required to describe it is about 10^{89} bits.

In summary, the Universe started out containing only 1,000 bits of information but now contains 10^{89} bits. That is an increase of 10^{86}, or 100 trillion trillion trillion trillion trillion trillion, times. It may not be obvious that this extraordinary increase in information is a puzzle, but it is. To understand why, it is necessary to understand something about the laws of physics.

The well-known laws of physics such as Newton's laws of

motion are recipes for predicting the future with 100 per cent certainty. For instance, if we know the location of the Moon today, by applying Newton's laws of motion and his law of gravity we can predict the location of the Moon tomorrow. Since knowing the location of the Moon yesterday is all that is needed to determine its location tomorrow, it follows that no new information is added. The location tomorrow is *contained* within the location today. And this is true of all the non-quantum, or 'deterministic', laws of physics. Since a single state of the system in the present completely determines a single state of the system in the future, no new information is created. In fact, deterministic laws are synonymous with the conservation of information.[7]

The relevance of this to the Universe is that its evolution is described by Einstein's theory of gravity – the general theory of relativity – which is a deterministic theory. In other words, general relativity, when applied to the whole Universe, describes how a particular state of the Universe evolves into another state at a later time. There is no change in the information content of the Universe.

What, then, are we to make of the fact that the pre-inflation Universe contained only 1,000 bits of information and today's Universe contains 10^{89}? In other words, there were only $2^{1,000}$ distinguishable ways the stuff of the Universe could have been arranged at the start yet there are a mind-numbing $2^{(10^{89})}$ ways it could be arranged now?[8] Since a non-quantum, or 'classical', theory like general relativity permits $2^{1,000}$ states at a particular time only to evolve into $2^{1,000}$ states at a later time, it can only mean that the complexity of the Universe today cannot have been determined at the beginning of time. 'Take the books on my bookshelf,' says Hsu. 'Why is my copy of *The Feynman Lectures in Physics* next to my copy of *A Brief History of Time*? The information argument is saying that the state of my bookshelf cannot be traced back to a unique state in the Big Bang. In fact, almost nothing in our Universe can be explained this way.'

Think of that leaf fluttering on a tree. What causes it to flutter? The wind, of course. But what causes the wind? Heat dumped into the atmosphere by sunlight. But what causes sunlight? Heat generated in the Sun's core by nuclear reactions . . . It may seem that such a chain of cause and effect can be followed all the way back to the birth of the Universe. But the fact that there is vastly more information in today's Universe than at the beginning is telling us that this is not true. Eventually, if the chain of cause and effect is followed back far enough, there will come an effect without a prior cause. Something which happened for no reason at all. An event which was utterly random.

And this is the clue to where all the information in today's Universe has come from. Randomness is synonymous with information. This is far from obvious. In fact, at first sight, it appears counter-intuitive. However, imagine there is a 100-digit number whose digits are random. The only way you can communicate it to someone else is to send all 100 digits. Contrast this with a number that consists of a string of a hundred 3s. You can communicate this by exploiting the pattern and simply saying '3 repeated 100 times'. This shows that a random number contains a lot of information, whereas a non-random number contains very little. A lot is redundant.

So what processes have been responsible for injecting the information/randomness into the Universe since the beginning of inflation? Hsu is in no doubt. 'Those processes can only be quantum processes,' he says. Quantum processes are non-deterministic. A unique state of a system in the past does not lead to a unique state in the future. The laws of quantum physics are not a recipe for predicting the future with 100 per cent certainty. They are a recipe for predicting myriad possible futures, each of which may happen with a particular probability. 'Things are happening in the world around us today because of countless bursts of randomness injected into the Universe since the Big Bang – because of the roll of a quantum dice,' says Hsu.[9]

Hsu believes that the principal process that injected randomness into the Universe was inflation itself. No one knows what drove it, although physicists often talk about an 'inflaton' field, some kind of 'stuff' which pervaded the Universe and whose repulsive gravity made the vacuum balloon enormously in size. During inflation, the Universe doubled in size, and doubled in size again more than 60 times over.[10]

At some point – and nobody knows why – inflation ran out of steam. The inflaton field decayed, leaving the vacuum as the normal vacuum we see around us today.[11] The key characteristic of this decay was that it was quantum – that is to say, random. This means that the inflaton decayed at slightly different times in different locations and dumped a different amount of energy in different locations. Since energy can neither be created nor destroyed, merely changed from one form into another, the energy of the inflaton manifested itself in other forms. Those forms were the mass-energy of subatomic particles and their energy of motion. In short, the decay of the inflaton made the matter of the Universe and simultaneously heated it up to a blisteringly high temperature. It created the 'hot' Big Bang.

So in the standard picture of cosmology, the Universe starts with just vacuum. The vacuum is in an unusually energetic state, which mushrooms in size wildly. The more vacuum that is created, the more vacuum energy there is. Inflation, as pointed out by many physicists, is the 'ultimate free lunch'. Finally, inflation ends and the energy of all the newly created vacuum heats up the Universe and creates the fireball of the Big Bang. Because the energy dumped into every location of the Universe by the decay of the inflaton was different, the temperature in each place was different too. 'The decay of the inflaton was like a random number generator, injecting a fantastic amount of randomness across the length and breadth of the Universe,' says Hsu. 'This quantum randomness is the reason my *The Feynman Lectures in Physics* is next to *A Brief History of Time*.'

The Universe has continued to expand since the end of inflation, creating more vacuum and more vacuum energy along with it, so you might think this would have injected more randomness into the Universe. However, the key difference between the expanding vacuum today and the inflationary vacuum is that the former has remained stable against decay. And it is only the decay of the vacuum – an inherently quantum process – that unleashes randomness into the wider Universe.[12]

Hsu does not believe, however, that the decay of the inflaton was the only process that injected randomness/information into the Universe. He believes that since the end of inflation, information has been continuously injected into the Universe by countless quantum events such as the random disintegration of atomic nuclei and the random emission of photons by atoms.

Recall, for instance, that the 'spin' of an electron may be clockwise or anticlockwise. Before it is recorded by some kind of detector, its spin is undefined. There is nothing to describe, no information. However, once the electron makes its mark on a detector – in the jargon, 'decoheres' – it is found to be either spinning clockwise or anticlockwise. There are two possible outcomes, which takes a single bit to describe. Where once there was no information, now there is some. Imagine ten spinning electrons in a row. If they impress themselves on some kind of detector, suddenly some are spinning clockwise and others anticlockwise. There are 2^{10} possible outcomes open to the ten electrons – more than 1,000 possibilities. So a whole load of information has been injected into the Universe just by these ten electrons registering their presence.

Just imagine how much information can be injected into the Universe by the trillions upon trillions of subatomic particles registering their presence. 'It's a tremendously powerful way of injecting information into the Universe,' says Hsu.

How is the increase of information in the Universe compatible with the remorseless increase of entropy in the Universe, as decreed by the second law of thermodynamics? Well, entropy and information turn out to be intimately connected: entropy $= e^{(\text{information})}$. The nineteenth-century physicists recognised that entropy increased because the number of disordered states open to atoms and their like overwhelmingly outnumbered the ordered states available. 'Although they did not know it, those states are in fact quantum states such as the state of an electron spinning in an atom,' says Hsu. 'Entropy increases because the disordered quantum states open to subatomic particles overwhelmingly outnumber the ordered ones. Everything fits.'

Einstein, as pointed out before, considered quantum processes – random and without cause – utterly abhorrent. However, according to Hsu, they are far from abhorrent.

They are absolutely essential. We owe our existence here today to quantum unpredictability. Look around you – at a rose, a newborn baby, a plane riding a vapour trail across the blue sky. We live in a world of boundless complexity. But all the complexity you see is merely the result of a long sequence of quantum coin tosses since the end of inflation. Like it or not, we live in a random reality.

Earth's Full, Go Home

How the fact there are no aliens on Earth is telling us either we are the first intelligence to arise or some unknown factor prevents the evolution of space-faring civilisations

'Sometimes I think we are alone, sometimes I think we are not. Either way, the thought is staggering.'

Buckminster Fuller

'I'm sure the Universe is full of intelligent life. It's just been too intelligent to come here.'

Arthur C. Clarke

One striking feature of the world is so obvious that, like the darkness of the sky at night, it is almost never remarked upon. It does not matter what country you live in, what continent you are on, where at all you are on the planet, there are no aliens. They are not loitering on street corners, coasting angelically through the clouds above your head or materialising and de-materialising like crew members of Star Trek'*s starship* Enterprise.

The fact that there are no aliens on Earth is widely believed to be telling us something profound about intelligent life in the Universe. Unlike the case with the other everyday observations in this book, however, no one is quite sure what that profound thing is.

Over the years, many people have realised that the lack of aliens on Earth is a deep puzzle. However, the person who articulated it in the most memorable way was the Italian physicist Enrico Fermi. One of the last physicists to combine the roles of front-rank theorist and experimentalist, not only did Fermi come up with a theory of radioactive beta decay, which predicted the existence of the ghost-like 'neutrino', but he constructed the first nuclear reactor – on an abandoned squash court under the west stand of the University of Chicago's Stagg Field. Fermi's 'nuclear pile', which went 'critical' on 2 December 1942, made the plutonium for one of the two atomic bombs dropped by America on Japan.[1] Those bombs were tested in the desert of New Mexico, and it was while visiting the bomb lab at Los Alamos in the summer of 1950 that Fermi made his memorable observation about extraterrestrials.

He was having lunch in the canteen with Herbert York, Emil Konopinski and Edward Teller, the 'father of the H-bomb'. The physicists had been discussing ETs because of a recent spate of newspaper reports of 'flying saucers'. Although the discussion had turned to more mundane subjects, Fermi had gone quiet, deep in thought. Suddenly, in the middle of the ensuing conversation, he blurted out: *'Where is everybody?'* The others around the table immediately knew what he was referring to – ETs. They also recognised that Fermi, a man with a reputation as a deep thinker, had articulated something important and profound.

Fermi was renowned for his back-of-the-envelope calculations. For instance, at the explosion of the first atomic bomb at Alamogordo in the New Mexico desert on 15 July 1945, he had dropped a scrap of paper from shoulder height and watched how it was deflected by the shock wave from the bomb. Knowing that Ground Zero was nine miles away, he estimated the energy of the blast – the equivalent of more than 10,000 tonnes of TNT.[2]

Implicit in Fermi's 'Where is everybody?' question was a similar back-of-the-envelope calculation. How long would it take a civilisation that developed star-faring capability to spread to every star system in our Milky Way?

Fermi never revealed the details of his reasoning. However, more likely than not he realised that the most efficient way to explore the Galaxy would be by means of self-reproducing space probes.[3] Such a probe, on arrival at a destination planetary system, would set about constructing two copies of itself from the raw materials found there. The two daughter probes would then fly off and, at the next planetary system, build two more copies. In this way, the probes would infect the Galaxy relatively rapidly, like bacteria spreading throughout a host.

Using plausible estimates for the speed of such probes and the time required to make copies, it was possible to estimate how long it would take to visit every star in the Milky Way. And the answer was surprisingly modest: between a few million and a few tens of millions of years. Since this was a mere fraction of the 10-billion-year lifespan of our Galaxy, one conclusion was unavoidable: if a star-faring race had arisen at any time in the history of our Galaxy, its space probes should be here on Earth today.[4] So, in Fermi's immortal words, 'Where is everybody?'

Really, Fermi's question is saying two things. Why don't we see any sign of ETs on Earth or in our Solar System? And, why do we see no sign of ETs when we look out at the Universe? For instance, why have we not seen some kind of technological artefact with our telescopes or picked up an ET signal?

Some people, of course, would maintain that there *is* evidence that ETs are here – unidentified flying objects. The majority of UFOs turn out to be rare atmospheric phenomena, high-flying weather balloons, the planet Venus seen under unusual conditions, and so on. However, a minority of sightings remain unexplained. This does not mean there is no natural explanation, simply that one has not been found. There is a serious objection to UFOs being extraterrestrial spacecraft, however, and that is the lack of material evidence. Despite claims of sightings of such craft – and in some instances even claims of direct contact with their occupants – nobody has come forward with a single alien artefact.

The American astronomer Carl Sagan voiced another objection to the idea that UFOs are alien spacecraft visiting Earth. He pointed out that such craft were suspiciously similar to modern high-tech aircraft, which often have the appearance of flying wings. To Sagan, it was like a Victorian looking up at the sky and seeing steam-powered flying machines. He called this tendency to see vehicles similar to present-day vehicles 'temporal chauvinism'. This flew in the face of an observation made by science-fiction writer Arthur C. Clarke: any sufficiently advanced civilisation – certainly one capable of crossing the vast distances between the stars – ought to be indistinguishable from magic. To Sagan, UFOs – though not necessarily hoaxes – were definitely in the eye of the beholder.

But what about alien signals? Surprisingly, radio telescopes over the years have picked up numerous broadcasts with the characteristic expected for an ET transmission – a signal spanning a narrow 'band' of frequencies much like a terrestrial radio station. The most famous of these is the 'Wow!' signal, registered on 15 August 1977 by the 79-metre 'Big Ear' telescope of Ohio State Radio Observatory. When astronomer Jerry Ehman saw the off-the-scale signal, he immediately scrawled 'Wow!' in the margin of the paper print-out. It is a name that has stuck. But despite reobservations of the sky in the direction of the 37-second-long transmission – the constellation of Sagittarius – the signal has never repeated.

This is characteristic of all the unusual signals that have been intercepted by radio telescopes, including 11 signals picked up in the late 1980s by the Planetary Society's Mega Channel ET Assay (META). It is possible they never repeat because they are from some transient source – maybe secret spy satellites in Earth orbit – or because they are the result of some freak electrical malfunction in the detecting equipment.

There does, however, remain the faint possibility that some of the unexplained signals are real. Joseph Lazio, of the National Research Council in Washington DC, and his colleagues have pointed out that there is electrically charged, or 'ionised', hydrogen gas drifting between the stars. It is distributed in an uneven, clumpy manner, and such irregularities are known to cause radio signals from distant astronomical objects such as 'pulsars' to fluctuate in brightness. Similar irregularities in the Earth's atmosphere cause stars to 'twinkle'. According to Lazio and his colleagues, such 'interstellar scintillation' can depress or boost a radio signal by as much as 20 times its normal strength. The scientists therefore suggest that the unusual signals could be ET broadcasts on the occasions when they are boosted.

Unfortunately, a signal subject to interstellar scintillation spends nearly all of its time depressed and is very rarely boosted. According to Lazio and his colleagues, it would require thousands or even tens of thousands of reobservations to see such a signal repeat. So far, however, no signals have been looked for in more than 100 follow-up surveys.

The scientists conclude that though the data is compatible with there being tens of thousands of ET civilisations in our Galaxy, it is equally compatible with there being none. ET may already have phoned Earth – only it is impossible to tell.

So if there is no evidence of ETs physically on Earth or in the Solar System and none that we are picking up intelligent signals from space, what should we conclude? Well, one possibility is that we have not yet looked hard enough for long enough. Certainly, this is the view of those actively involved in the Search for Extraterrestrial Intelligence, or SETI. They point out that so far only a tiny fraction of the 200 billion or so stars in our Galaxy have been targeted for radio signals.

Nevertheless, according to Seth Shostak, of the SETI Institute in Mountain View, California, we are now observing target stars at a furious rate. He believes that to stumble on the first alien civilisation, we will need to observe a few million stars, a number that will be reached by 2015. He will be surprised, he says, if SETI does not meet with success by that date. Shostak's tacit assumption, generally accepted in the SETI community, is that between 10,000 and a million alien civilisations are currently broadcasting radio signals in our Galaxy. Sceptics consider the estimate at best optimistic and at worst wishful thinking. SETI scientists lean towards the view of the sixteenth-century Italian philosopher Giordano Bruno, who, in his book *On the Infinite Universe and Worlds*, declared: 'There are innumerable suns, and an infinite number of earths revolve around those suns, just as the seven we can observe revolve around this sun which is close to us.'

The idea that we have not looked for ET signals hard enough or long enough could equally well explain why we have seen no sign of alien visitation in our Solar System.

Recall that in Stanley Kubrick and Arthur C. Clarke's *2001: A Space Odyssey* an alien artefact is found on the Moon. Left there 3 million years before as a kind of 'baby alarm' to alert its makers if intelligence ever arose on Earth, its presence is not obvious because it is buried in the Tycho crater, deep beneath the lunar dust. It is plausible that an alien artefact might be left somewhere even less obvious. As physicist Stephen Webb points out, there are 50 billion billion billion cubic miles of space within a sphere that encloses the orbit of Pluto, and the Solar System extends to the Oort Cloud of comets, far beyond Pluto. 'The chances of finding a small alien artefact by accident are essentially zero,' he says.

Of course, we may have seen no signs of aliens and picked up no alien signals not because there are still so many places left to look but because we are using the wrong search strategy. There is a tendency for SETI scientists to target Sun-like stars for the obvious and sensible reason that we know of one such star around which life has definitely arisen: the Sun. But perhaps ET civilisations, because of their rapidly mushrooming energy needs, move from relatively cool stars like the Sun, of 'spectral type' G and K, to super-hot O and B stars. Such stars are rarely targeted by SETI searches because they burn up quickly and would incinerate any planets that happened to have formed.

Then again, perhaps most life is not to be found around stars at all. In an attempt to widen the debate about life in the Galaxy, planetary scientist David Stevenson, of the California Institute of Technology in Pasadena, has suggested that life might most commonly be found on 'interstellar planets', sunless orphans wandering through the dark deep freeze between the stars. As evidence for his idea, he points to computer simulations of the formation of our

Solar System, which invariably show about ten Earth-mass planets forming. Close encounters with embryonic giant planets like Jupiter quickly cause the majority to be ejected into interstellar space.

Naively, it might seem that a planet without a sun to warm it and provide energy would be a dead loss for life. However, Stevenson points out that such planets will be swathed in a thick embryonic atmosphere of molecular hydrogen gas, the raw material from which stars are born, and this atmosphere could be anything from 100 to 10,000 times denser than the Earth's atmosphere. Crucially, under such high-density conditions, all gases become 'greenhouse' gases, trapping heat with the effectiveness of a planet-wide duvet. According to Stevenson, with such good insulation the heat coming from radioactive rocks – the same heat that to this day keeps the Earth's interior molten – could keep the surface of such a planet warm enough for liquid water to exist for at least 10 billion years, twice as long as the current age of the Earth. Remarkably, interstellar planets might be the most likely places in the Galaxy to find life, most probably primitive microorganisms but – who knows? – maybe intelligence as well.

But even if we are not looking in the wrong place for ETs, another possible flaw in our search strategy might be that we are listening at the wrong radio frequencies. Searches for intelligent radio signals, for instance, tend to focus on a small number of special radio frequencies. These are the frequencies of radio waves emitted naturally by atoms and molecules commonly found drifting on the currents of space. ETs will know about these frequencies, goes the argument, and they will know that other technological civilisations will know them too. The 'Wow!' signal, for instance, was picked up at 1,420 megahertz (MHz), the frequency at which the hydrogen atoms floating in interstellar space broadcast to all and sundry like miniature radio stations.

It is entirely possible that ETs are broadcasting at radio frequencies which we have not anticipated. Although searches have also been conducted for extraterrestrial intelligence signalling with visible light, the same logic applies. ETs might be using optical frequencies we have not yet tried.

But maybe our search strategy is even more flawed than this. Aliens, rather than using different radio waves or visible light to signal, may be using an entirely different communication medium altogether. For instance, they might be signalling with ghostly neutrinos or with gravitational waves – ripples in the fabric of space–time – or using some other communication mechanism we cannot begin to imagine. By looking for radio or optical signals, and expecting ETs to communicate in pretty much the way twenty-first-century humans do, we may be guilty of Sagan's temporal chauvinism.

In his book *The Cosmic Connection*, Sagan describes tribes that live in deep valleys in New Guinea and which use drums to communicate with people in adjacent valleys. When asked how an advanced tribe might communicate, tribesmen say: 'By using a bigger drum.' The irony is that all the while, the babble of global radio traffic fills the air. In the same way, we might be completely oblivious to the babble of Galactic communication surging through the vacuum all around the Earth. Arthur C. Clarke had a similar thought. In *Odyssey*, an authorised biography, Neil McAleer quotes Clarke saying:

> The fact that we have not yet found the slightest evidence for life – much less intelligence – beyond this Earth does not surprise or disappoint me in the least. Our technology

must still be laughably primitive, we may be like jungle savages listening for the throbbing of tom-toms while the ether around them carries more words per second than they could utter in a lifetime.

The truth, of course, is that we have little choice in our search strategy. Either we abandon the search for extraterrestrial intelligence altogether or we listen with the means currently at our disposal and keep our fingers crossed.

But it is possible that it is not only ET communications that are unrecognisable. Alien artefacts – in the Solar System or elsewhere – may also be unrecognisable. After all, ETs could be millions or even billions of years ahead of us technologically. Does an ant recognise that it is crawling over a paving stone in a city connected to other cities by land, sea and air? Does a bacterium? Once again, Arthur C. Clarke's remark is apt: a sufficiently advanced civilisation will be indistinguishable from magic.

But maybe we should take the fact that we have not seen any evidence of ETs at face value. The question then is: why have they not come here or left any discernible sign of their presence in the heavens?

Could it be that it is physically impossible to cross interstellar space – that the distances are simply too vast? This seems unlikely. Even today we can envision means of travelling to the stars at, say, 1 per cent of the speed of light. Although it is beyond our present-day capabilities, it seems unlikely it will always be.

If there is no physical barrier to travelling between the stars, then perhaps ETs find something better to do with their time than explore the Galaxy or beam radio messages to all and sundry. Perhaps they become preoccupied with art or introspection and decide to stay at home and ignore the wider Universe. Or perhaps they find computer-generated artificial realities more seductive than the messy real world. Perhaps they self-destruct, snuffed out by nuclear war or global warming or any number of other possible environmental catastrophes. Or maybe ETs adhere to something like the *Star Trek* prime directive on not interfering with the emerging civilisations. According to this 'zoo hypothesis', our Solar System is cordoned off. We can expect to be left alone on our 'nursery world' until, one day, we develop starfaring capability and sail out into interstellar space to be welcomed as new members of the Galactic Club.

One of the problems with these kinds of explanations would appear to be that they require us to speculate on the motivations of ET civilisations far in advance of our own. Since we cannot know their motivations, we are on shaky ground, to say the least. Fortunately, however, such explanations of why we see no sign of ETs can be dismissed regardless of ET motivations. As physicist Michael Hart of the National Center for Atmospheric Research in Boulder, Colorado, pointed out in a classic 1974 paper, we can expect ETs to have a variety of motivations. Sure, some may blow themselves up. Sure, there may be stay-at-home ETs. Maybe even the majority will stay at home. But just as a small minority of the world's population has gone out to explore the four corners of the globe while the majority has stayed at home, there will always be exceptions that buck the couch-potato ET trend. And while most ETs might respect a zoo-like quarantine of the Solar System, would all of them? A policy of non-interference might be difficult to police over millions or even billions of years – and there is no evidence the Earth has been interfered with during its 4.55-billion-year lifetime.

According to Hart, all the explanations of the absence of ETs based on their motivation are undermined by the same thing: inevitably, there will be exceptions who have different motivations to the rest or who ignore bans on visiting the Solar System – rogue ETs who nevertheless come to Earth.

So might it be possible to explain the non-appearance of ETs on Earth in some way that does not rely on ET motivations? Might there be some compelling reason not to go from star to star, exploring the Galaxy? One possibility is that it is too dangerous. 'Any creatures out there may be malevolent or hungry,' pointed out radio astronomer Sir Martin Ryle at the University of Cambridge on hearing, on 16 November 1974, that astronomers at the giant radio dish in Arecibo, Puerto Rico, had announced our presence to the cosmos by broadcasting the first interstellar radio message in the direction of the globular star cluster M13 about 25,000 light years away. Perhaps, as Ryle worried, there is a xenophobic race out there that wipes out other civilisations soon after they make their presence known. Maybe self-reproducing space probes accrue mutations with every new generation just as living things do until, eventually, a mutation leads to a murderous species which sees as its mission the wiping out of all competing life forms. Science-fiction writer Fred Saberhagen envisioned just such a race of doomsday machines, which he called 'Berserkers'. If Berserkers are out there, it might be best to sit quiet and listen rather than broadcast our existence to the four corners of the Galaxy. The trouble is, it may already be too late since our TV broadcasts, travelling at the speed of light, have already reached stars more than 70 light years away.

A non-Berserker reason why ETs might not want to explore the Galaxy is that there is some place far more attractive, far more seductive to visit. Astronomers have found that there is six times as much invisible, or 'dark', matter in the Universe as normal matter – the atomic stuff of which you and me, the stars and galaxies are composed. We know of its existence only by the gravitational tug it exerts on the visible material. Even the dark matter is outweighed by the mysterious dark energy, with the two invisible cosmic components together accounting for 96 per cent of the mass-energy of the Universe. Could it be that the Universe is chock-a-block full of dark-matter stars, dark-matter planets and dark-matter life? Might it be that all the activity, the cosmic commerce, is going on in the dark matter, that the dark, not the normal, matter is where it's at? As *New Scientist* reader Simon Williams observed in the magazine's letters page in 2003: 'We are told that only 4 per cent of the Universe is made of atomic matter that we can observe, and that the remaining 96 per cent is missing. Perhaps someone or something else is sitting out there trying to figure out where the missing 4 per cent is.'

Then, in addition to dark matter, there is a strong suspicion among physicists that there may be other dimensions. According to 'string theory', which views the ultimate building blocks of matter as tiny vibrating strings of mass-energy, there are six space dimensions in addition to the three of space and one of time with which we are familiar. The extra dimensions are thought to be 'rolled up' far smaller than an atom, but there are ways theorists believe they could be big and still have escaped notice. Could ETs find more interest, more activity, in other dimensions? Or maybe they escape into other, more interesting universes down short-cuts in space–time called 'wormholes', which are permitted to exist by Einstein's general theory of relativity. Maybe even – God forbid – we are simply not interesting enough and aliens treat us with monumental indifference. 'How would it be if we discovered that aliens only stopped by Earth to let their kids take a leak?' wondered American chat-show host Jay Leno.

But, of course, it is not just that ETs are not here on Earth, that they do not appear to have explored the Galaxy. We see no sign of them either. Even if they do not signal, we might expect to see signs of their presence. As pointed out, our TV broadcasts have already leaked out into neighbouring space. On planets around stars 70 light years away it should be possible to pick up Nazi broadcasts from the 1930s. And it is not just the 'electromagnetic radiation' of TV. There are military radars and power lines and the whole electrical paraphernalia of our technological civilisation. Surely if our stuff is leaking out into space we should also expect to see the same kind of electromagnetic leakage from advanced technological civilisations? It does not matter whether ETs, for whatever reason, decide to stay at home gazing at their alien navels, they should inadvertently reveal their presence. But they have not.

Of course, absence of evidence is not necessarily evidence of absence. As pointed out already, we may not yet have looked hard enough or long enough. However, it looks like there is no one out there. Some scientists say we should not be swayed by emotion, by our heart telling us that surely we cannot be totally alone in the Universe. Instead, we should take a cold, hard look at the facts and draw the logical conclusion without flinching. The fact is, we see not the slightest piece of evidence for ETs. And the logical conclusion to draw is that this is because they do not exist. We are alone, the first intelligence to arise in our Galaxy.

Such a conclusion, though logical, not only flies in the face of our desire not to be alone in the cosmos but collides with the idea that there is nothing special about our place in the Universe. In the sixteenth century, the Polish astronomer Nicolaus Copernicus told us that the Earth orbited the Sun along with all the other planets and so had no special position in the Solar System. In the nineteenth century, Charles Darwin taught us that humans are just another animal species, the product of evolution by natural selection. Then, in the twentieth century, we learnt that the Milky Way was but one among another 100 billion galaxies and had no special position in the Universe.

To accept that we are special – the only intelligent life in the Galaxy (possibly even the whole Universe) – flies in the face of the 'Copernican Principle', also known as the principle of mediocrity. It smacks almost of religion, in which the human race is assumed to be a special project of a creator, despite the fact that the Earth is but one planet among possibly trillions upon trillions of others. But difficult as it is to accept that we are alone, it does appear to be the logical conclusion to draw from what American science-fiction writer David Brin calls 'The Great Silence'.

But how could it be that we are the first intelligence to arise, at least in the Milky Way? Well, if we are, it must mean that the emergence of intelligent life is a spectacularly unlikely occurrence. Perhaps it requires a whole chain of accidents, each of low probability.

One low-probability event appears to be the formation of a giant satellite. The Earth's moon has been essential for life on Earth. For one thing, it has stabilised the climate over billions of years. This is because the Earth, as it spins, has a tendency to wobble and fall over, much like a top. Such wild variations in the spin axis of the Earth would result in wild fluctuations in its climate. In fact, this is exactly what happens on Mars, which has only two minuscule moons. As Elton John sang in 'Rocket Man': 'Mars ain't no place to raise your kids.' However, whenever the Earth tips over too far, the gravity of the Moon pulls it back. The Moon can do this only because it is a quarter of the diameter of the Earth and so has an appreciable gravitational pull. In fact, the Earth–Moon system is effectively a double planet.

But look around the rest of the Solar System. There are hundreds of moons, but only one other – Pluto's Charon – is comparable in size to its parent planet.[5] The rarity of such moons is believed to be because of the unusual circumstances required for their formation. Shortly after its birth, the Earth is believed to have been struck by a Mars-mass body. According to this 'Big Splash' theory, the impact turned the exterior of the Earth molten, splashing some of it into space to congeal as the Moon. This may have been an unlikely event. So if big moons are essential for life-supporting planets, then this greatly reduces the number of possible sites for life in the Galaxy.

In addition to the accident of the Moon, there is the fact that the Earth has both water and land. Although whales and dolphins are known to be as bright as the higher apes, and on the very threshold of human intelligence, it was only on land that the step to human intelligence was taken. Perhaps it was only there that evolutionary pressures were sufficient to create a hand capable of dexterous manipulation of materials to make tools and weapons. If this is so, then the step to human-level intelligence would not be taken on planets entirely covered with water. And since the Earth is itself almost entirely covered with water, it is conceivable that this might be the case for most life-supporting worlds. A galaxy of dolphin-like creatures might be a silent galaxy.

But in addition to a big moon and some dry land, biologists can identify at least a dozen hurdles that life had to overcome on the road to producing human beings. A key one, however, appears to be the step from single-celled to multicellular organisms, which happened only about 700 million years ago, nigh on 3 billion years after the emergence of life on Earth. Since single-celled organisms such as bacteria can reproduce in a matter of hours, we are talking about many trillions of generations before cells hit on the idea of clubbing together to form higher organisms such as ants and human beings. Could this be the hurdle that has prevented life evolving to the level of human intelligence or beyond everywhere else in the Galaxy?

The physicist Brandon Carter, one-time office mate of Stephen Hawking, has provided an ingenious mathematical argument stating that there were five low-probability, 'hard' steps on the evolutionary road to human technological civilisation. Step 1 was the advent of the first bacteria, or 'prokaryotes'; step 2, the complex cells with nuclei, or 'eukaryotes'; step 3, multicellular life; step 4, intelligence; and step 5, human civilisation. Each step took roughly 800 million years. There is also the possibility that life started on Mars, which is smaller than the Earth and so cooled from its initial molten state more quickly, before being transferred to Earth on board a meteorite. If so, we may be talking about six hard steps. The upshot is that if Carter is right about the five – possibly six – low-probability steps on the road to human beings, technological civilisation is likely to be extremely rare in our Galaxy.

The obvious question is then: why did it happen here on

Earth first? But, of course, it had to happen somewhere first. Why not here? Certainly, the Great Silence is utterly compatible with the human race being the first technological civilisation in our Galaxy.

At this point, it is worth pointing out that if the step from single cells to multicellular organisms took an unimaginably long time, the step from non-life to life – from inert chemicals to the first bacterial cells – was remarkably rapid on Earth. There is evidence of primitive life on Earth 3.8 billion years ago, within 800 million years of the planet's birth, which means it had to be around even earlier. In fact, it must have arisen almost at the moment the newborn Earth had cooled enough from its molten state for life to be possible. Despite this, however, it has proved impossible – even after more than half a century of effort – to turn non-life into life in the laboratory.[6]

Some say the only way to resolve the paradox of life's rapid emergence on Earth and its difficulty in getting started is to say that the Earth was 'seeded' from space – Mars being an uncontroversial possibility – with the most primitive microorganisms. If so, we are aliens – or at least we arose from aliens. This idea of 'panspermia', though dating back to the late nineteenth century, was championed in the twentieth century by the late Fred Hoyle and Chandra Wickramasinghe. According to the two astronomers, life is extremely difficult to get started, but once it starts – perhaps in only one place in the Galaxy – there is an efficient mechanism to spread it from star to star, planet to planet. The proposed mechanism involves interstellar clouds and comets and a great cosmic cycle that continually ferries bacteria between the depths of space and planetary surfaces and back again.[7]

Nobody knows whether panspermia is true. But it remains a viable alternative to the view that life arose on Earth in splendid isolation. Certainly, if it is true, then primitive life is widespread – a cosmic rather than a planetary phenomenon. Of course, if there is a big hurdle on the evolutionary road to intelligence – perhaps at the step from single-celled to multicellular organisms – panspermia is still no consolation to us. The Universe could very well be teeming with bacteria but bereft of any complex or intelligent life. We will still have no one to talk to.

So is the depressing answer to Fermi's question – 'Where is everybody?' – nowhere, because we are the first? Not necessarily. According to one man, intelligent life is inevitable. The trouble is, there is a snag. Intelligent life is inevitable, but we will never, ever find it – at least not by looking out into the Galaxy.

As evidence, the British physicist Stephen Wolfram points to our communication signals. In order to squeeze more and more information into them – be they mobile-phone conversations or computer data – we remove all redundancy or pattern.[8] If anything in a signal repeats, then clearly it can be excised. But this process of removing any pattern from a signal makes it look more and more random – in fact, pretty much like the random radio 'noise' that rains down on Earth from stars and interstellar gas clouds. According to Wolfram, if someone beamed our own twenty-first-century communication signals at us from space we would have a hard job determining whether they were artificial or natural. So what chance do we have of distinguishing an ET communication from the general background radio static of the cosmos?

As it is for ET signals, so it is for ET artefacts. According to Wolfram, they too will be unrecognisable, though this is more difficult to see. He uses the example of a train station seen from so high above that the details of the trains are invisible. What would tell you that the train station is the product of technology is the regularity with which trains arrive and leave. However, Wolfram maintains that such a pattern will be totally absent from transport systems of the future. He believes they will use lots of small cabs summoned on demand. Such a system, coordinated by computers, would look entirely random viewed from high up. It would look far more like a natural artefact.

At present, says Wolfram, it is easy to distinguish a technological artefact such as a car from a natural object such as a tree. The tree is far more complicated. But, says Wolfram, this is simply because our technological artefacts are primitive. As they become more complex – with computer processors enabling them to make moment-by-moment decisions – they will begin to look just as complex as trees and people and stars. So what chance do we have of distinguishing an ET artefact from a natural celestial object? Essentially none, says Wolfram.

If Wolfram is right and ETs are out there but we cannot recognise them, either in their communications or their artefacts, then of course they could be here in the Solar System, neatly solving the Fermi paradox. That tree on the corner of your street could be an ET. Wolfram thinks this is unlikely, however. In fact, he thinks he has a cast-iron reason why ETs will not want to travel to Earth – or anywhere else for that matter. And it is all to do with computers.

This next bit is difficult to swallow. Wolfram thinks he has found nature's big secret – how it generates the complexity of the world, everything from a rhododendron to a tree to a barred spiral galaxy. It does it, he believes, by applying simple rules over and over again; by running simple computer programs, if you like. He came to this remarkable conclusion in the early 1980s, when he discovered that the simplest kind of computer program – known as a cellular automaton – can generate infinite complexity if its output is repeatedly fed back in as its input.

Crucially, Wolfram has found evidence that the kind of computer program that produces endless complexity can be implemented not just in systems of biological molecules but in all sorts of physical systems – chaotic gas clouds, systems of subatomic particles and so on. He concludes that all over the Universe, life – though definitely not life as we know it – will spring up spontaneously. It is a fundamental feature of matter.

This would appear to be good news. However, not only will ET life be unrecognisable, for the reasons already expounded, it will not want to come to Earth, according to Wolfram. The reason is subtle.

In Wolfram's view, everything in the Universe is the product of a computer program. In fact, he imagines an abstract cyber-universe of all conceivable computer programs, all the way from the simplest up to the most complex. This 'computational universe' contains everything from the Apple Macintosh operating system to a program for creating a faster-than-light starship. The existence of this computational universe is the crucial thing, for if any ET civilisation contacted us, says Wolfram, what could it trade with us? In his view, all it could say is, 'Here are some useful programs we have found in the computational universe. What have you found?' But the reality is it would be easier and more efficient to stay at home and use a computer to search the computational universe for useful programs rather than try to get the same information by hunting for ETs to talk to among the 200 billion or so stars in the Galaxy. 'It's a simple numbers game,' says Wolfram.

Perhaps you think such a stay-at-home ET would miss out on the pleasure of meeting us and learning about our civilisation at first-hand – or anyone else's for that matter. But think again. 'Remember, everything is generated by computer program – and that includes you and me,' says Wolfram. 'Someone halfway across the Galaxy could have found the computer program for you and be conversing with you at this very moment.'

If you do not like Wolfram's rather unorthodox explanation of the Fermi paradox, then you are left essentially with two plausible options: either there is some murderous race out there in the Galaxy – in which case perhaps we should be cautious about overtly announcing our presence – or we are the first intelligence to have arisen and are therefore utterly alone in the Milky Way. Then again, there could still be some explanation nobody has yet thought of.

Maybe someone has stuck a sign at the edge of the Solar System saying 'Earth's full, go home.' In any case, it seems right to end this book on everyday observations which tell us profound things about the Universe with an everyday observation – the lack of ETs in our midst – for which we do not yet know what the profound thing is. Speculate away. Your guess is likely to be as good as mine. Fermi's reaction to a lecture he once attended encapsulates the situation nicely, so it seems fitting to leave the last word to him: 'Before I came here I was confused about this subject. Having listened to your lecture I am still confused. But on a higher level.'

Acknowledgements

My thanks to the following people who either helped me directly, inspired me or simply encouraged me during the writing of this book: Karen, Henry Volans, Felicity Bryan, Neil Belton, Larry Schulman, Jim Hartle, Stephen Hsu, Michaela Massimi, David Arnett, Ed Harrison, A. C. Grayling, Stephen Hawking, Adrian Mitchell, Freeman Dyson, Ian Bahrami, Simon Singh, Brian May, Sara Menguc, Al Jones, Tania Monteiro, Brian Clegg, Clare Dudman, Alex Holroyd, John Grindrod, Sarah Savitt, Miles Poynton, Michele Topham, Shanaz Mirza, Patrick O'Halloran, Andy and Candy Coghlan, Jeremy Webb, Valerie Jamieson, Roger Highfield, Bobbie Derbyshire, Alexander Gordon Smith, Alom Shaha, Steve Hedges, Sue O'Malley, David Hough, Pam Young, Hazel Muir, Stuart and Nikki Clark, Spencer Bright, Chrissy Iley, Karen Gunnell, Jo Gunnell, Pat and Brian Chilver, Stella Barlow, Barbara Pell and David Brewin, Julia and Bill Bateson, Anne and Martin Ursell, Barbara Kiser, Diane and Peter and Ciaran and Lucy Tomlin, Eric Gourlay, Paul Brandford, Helen and Steve, Lucy, Chris, Helen and Olivia, Nuala, Naveen and Ibrahim, Mary, Axel and Simona, Carol and Bill.

Notes

CHAPTER 1

1. Actually, rather than visible light, Compton used X-rays. This ultra-high-energy light had so much oomph that it easily knocked electrons out of atoms. To all intents and purposes, they reacted like free-floating electrons rather than ones tied to an atomic nucleus.

2. Remarkably, relativity could have been a natural and unsurprising outgrowth of *sixteenth-century* physics. As several people have realised since Einstein, relativity is actually an unavoidable consequence of two things. One is that the laws of physics look the same whatever your state of motion, as long as that motion is at constant velocity. For instance, a ball thrown between two people follows the same shaped trajectory whether they are standing in a field or on a train travelling at 100 kilometres per hour. And the second thing is that the laws of physics look the same no matter what your orientation in 3D space. It is not necessary to assume anything about the speed of light, as Einstein did. Galileo could have discovered relativity. See 'The Theory of Relativity – Galileo's Child' by Mitchell Feigenbaum (http://xxx.lanl.gov/abs/0806.1234).

3. In fact, it has more than stood the test of time since it turns out that it is not only matter that is grainy but everything. This is the meaning of the word 'quantum' in quantum theory. A quantum is an indivisible grain of something. Matter comes in quanta. So does energy, electric charge, time, and so on. We live in a fundamentally grainy world.

4. It is always possible there is a deeper level of reality beneath quantum theory and that the probability of things happening is determined by factors operating at this fundamental level, just as the roll of a dice is determined by environmental factors. This

219

possibility continues to be explored by some scientists, including the English physicist Antony Valentini and the Dutch Nobel Prize-winner Gerard t'Hooft. However, they are in a minority. The theory appears to work perfectly if the unpredictability is indeed nature's fundamental, irreducible bedrock, so most physicists see no compelling reason to look any deeper.

5. Another irony is that, in 1900, the year Planck proposed the quantum, Lord Kelvin, one of the greatest physicists of his day, surveyed the achievements of his contemporaries and declared: 'There is nothing new to be discovered in physics now. All that remains is more and more precise measurement.' How wrong he was.

6. Young's double-slit experiment is one of the pivotal experiments in the history of science. Today, however, you can prove that light is a wave with a £1 laser pointer and a £2 metal ruler. Simply shine the laser at a very shallow angle along the metal ruler so that its narrow beam spreads out enough to illuminate several of the most closely spaced gradations on the ruler. Each of the gradations will act as a secondary source of concentric light waves which, as they spread through space, will pass through each other. Where they reinforce, they will create bright spots, and these will show up if, for instance, there is a convenient white wall in the path of the light. Strictly speaking, the spots are a result of 'diffraction', a phenomenon closely related to interference, but an undeniable characteristic of waves nevertheless.

Further reading:
The Magic Furnace by Marcus Chown (Vintage, 2000).

CHAPTER 2

1. See Chapter 1.

2. Troublesome exceptions are elements such as chlorine, which weighs in at 35.5 times the weight of hydrogen. Prout did not know that chlorine comes in several types, each of which, individually, weighs an exact multiple of hydrogen's weight but whose average is 35.5 times that of hydrogen.

3. Actually, on emission from radium, alpha rays were merely the cores, or 'nuclei', of helium atoms, but that is getting ahead of the story. But by the time Rutherford detected them, they had combined with electrons to make helium atoms.

4. In fact, there is a third type of ray that can be emitted by a radioac-

tive substance. A 'gamma ray' is an ultra-high-energy form of light.

5. The term 'nucleus' was not used until 1912.

6. See Chapter 1.

7. The dark energy is invisible and fills all of space, and its repulsive gravity is speeding up the expansion of the Universe. Its energy density is a whopping 1 followed by 120 zeros smaller than predicted by quantum theory, our current best description of ultimate reality.

8. A word on this 'scientific notation' which is so commonly used by physicists. 10^{34} means 10 multiplied by itself 34 times. And 10^{-34} means 1/(10 multiplied by itself 34 times). So, for instance, $10^5 = 10 \times 10 \times 10 \times 10 \times 10 = 100,000$. And $10^{-3} = 1/(10 \times 10 \times 10) = 1/1000 = 0.001$.

9. De Broglie thought matter waves were really waves of matter. But recall that the wave associated with a particle like an electron is actually more abstract than that. It's a probability wave, which spreads according to the Schrödinger equation and whose height at any location – strictly speaking, the square of the height – is related to the chance, or probability, of finding the particle there.

10. Another popular explanation is that there is an infinite number of parallel realities stacked like the pages of a never-ending book. According to this 'Many Worlds' picture, when a particle is in a superposition which corresponds to being in two places at once, it is not actually at two places at once in one reality; it is at one place in one reality and another place in a neighbouring reality. In this view, a particle goes through only one slit in the opaque screen, but it interferes with a particle that went through the other slit *in a neighbouring reality*.

11. Here we are still talking about the 'act of observation', or interaction of the bullet with the wall, imparting some sideways jitter to the bullet. In other words, we are saying the uncertainty is not intrinsic to the bullet but caused by the act of observation. In fact, it is intrinsic. A better/complementary explanation is decoherence.

Further reading:
'The Sun Likes Me', *Heart on the Left* by Adrian Mitchell, p. 194.

CHAPTER 3

1. Mass here means 'rest mass'. Some particles, such as photons, have no rest mass. They are born travelling at the speed of light and cannot exist at rest with respect to anything or anyone.

2. See Chapter 2.

3. Of course, latitude and longitude are used to specify a *location* on the surface of a globe. But this is possible only because every point on the surface is at the same distance from the centre of the Earth.

4. What all this is emphasising is that quantum theory is just a theory about what we can know or measure – which, really, is what any scientific theory should be. If we can know only the outcome of an event such as the interaction between two identical particles, we have no right to ask how that outcome came about. In fact, the question is not a legitimate scientific question. It has no meaning. As Niels Bohr said: 'It is wrong to think that the task of physics is to find out how nature is. Physics concerns what we can say about nature.'

5. Which I did not answer in my book *Quantum Theory Cannot Hurt You* (Faber, 2008).

6. Actually, all particles with half-integer spin – ½, ³⁄₂, ⁵⁄₂, and so on – are fermions, and all particles with integer spin – 0, 1, 2, and so on – are bosons.

7. The proof that spin-½ particles (or, more generally, particles with half-integer spin) obey the Pauli exclusion principle was not straightforward. It was not until 1940 – 16 years after he had discovered the exclusion principle – that Pauli proved the so-called spin-statistics theorem.

8. It is the alignment of an electron's spin in a magnetic field that gave the first hint of the existence of spin. When an electron in an atom jumps between one state and another – a so-called quantum jump – it spits out or absorbs light of energy equal to the difference in energy of the states. However, in a magnetic field, the light could have two slightly different energies – on either side of the expected energy. The explanation is that the electron's spin can be aligned with the magnetic field or against it, and that each of the orientations corresponds to a slightly different energy. Peculiarly, Pauli discovered the exclusion principle *before* the discovery of spin by the Dutch-American physicists Samuel Goudsmit and George Uhlenbeck in 1925/26. Although Pauli did not know of spin, however, he nevertheless knew of the twofoldness, or *zweideutigkeit*, of the energy states of an electron in a magnetic field. See *Pauli's Exclusion Principle: The Origin and Validation of a Scientific Principle* by Michela Massimi (Cambridge University Press, 2005).

9. See Chapter 2.

Further reading:

QED: The Strange Theory of Light and Matter by Richard Feynman (Penguin, 1990). This is the most brilliantly simple book on the most successful physical theory ever devised. I cannot praise it enough. Feynman was not only a genius physicist, he was a genius populariser too.

A Brief History of Time by Stephen Hawking (Bantam, 1995).

The Great Physicists from Galileo to Einstein by George Gamow (Dover Publications, New York, 1988).

CHAPTER 4

1. The only real influence the composition of a body has is on how easily heat escapes from it. This depends on the number of free electrons, since free electrons turn out to be good at 'scattering', or redirecting, radiant heat, hindering its progress outwards from the centre. An object predominantly made of hydrogen has a maximum of one free electron per atomic nucleus to dam up internal heat, whereas an object made of heavier atoms has more.

2. Anaxagoras may have been this precise because he was the first to realise that the Moon is opaque and so casts a shadow on the Earth when it passes in front of the Sun. He was able to gauge the extent of the shadow from eye-witness reports, principally from sailors, during the annular eclipse of 478 BC. The shadow covered the Peloponnesian peninsula. So Anaxagoras concluded that the Moon was 'as large as the Peloponnese' and the Sun therefore 'a little larger than Greece'.

3. See Chapter 3.

4. Initially, Herschel christened Uranus 'George' in honour of England's monarch, George III. Not many people know this.

5. This estimate was made in 1658 by the Irish archbishop James Ussher.

6. The more sophisticated explanation of this uses the law of conservation of energy. In the act of pumping, the energy of motion of a piston is transformed into the heat energy of the air behind the piston – in other words, the random, frenzied motion of the air molecules.

7. See Chapter 2.

8. If the decay of an atom is unpredictable, then it implies that if you watch it for, say, ten minutes, then another ten minutes, and another ten, and so on, the atom will have the same chance of decaying in each interval. If it did not have the same chance and,

say, was more likely to decay between 30 and 40 minutes than in any other interval, then clearly its behaviour would no longer be unpredictable – you would know it would be more likely to decay between 30 and 40 minutes. Now, say we have a sample containing a large number of radioactive atoms. And say the chance of it decaying in the first ten minutes is 1/2. After ten minutes, therefore, half the atoms will be undecayed. After 20 minutes, a half of what are left, which is a quarter, and so on. This trivial example shows how an equal chance of decay in each ten-minute interval leads to a decay law with a half-life of ten minutes. What is not so trivial is to realise that even if the chance of an atom decaying in any particular interval was 1/10 or 1/63 or 0.000023, the same kind of radioactive decay law characterised by a particular half-life would still result.

9. I tried to rectify the injustice that had been done to Cecilia Payne by writing an article about her work in 2003 in the magazine *New Scientist* (http://newscientist.com/article/mg18024205.300-the-star-who-unravelled-the-sun.html). Unfortunately, there was a cock-up at the printers and, when the article appeared, the photograph that accompanied the article, instead of showing Cecilia Payne, showed only the corner of her hat. Well, I tried. But perhaps some people are cursed, destined never to get the credit they deserve.

10. In fact, all that Atkinson and Houtermans did was turn on its head an idea originated by their colleague, George Gamow. He was the first to apply quantum theory to the atomic nucleus in an attempt to explain radioactive alpha decay, in which a nucleus of helium is ejected at ultra-high speed from an unstable heavy nucleus such as radium. The problem is that alpha particles have insufficient energy to escape their nuclear prison – they are trapped down a mine shaft – yet they still do escape, appearing spontaneously on the lower slopes of the nuclear hillside. The key to their Houdini-like behaviour, Gamow realised, is their quantum nature. Though they have insufficient energy to climb to the lip of the mine shaft, their spread-out waviness enables them to 'tunnel' through the hillside to freedom.

CHAPTER 5

1. It would turn out, however, that the lightest nuclei such as deuterium – heavy hydrogen – and helium *were* made in the fireball of

the Big Bang. In fact, after the ten-minute fury of nuclear reactions, roughly 10 per cent of the nuclei have become helium, a proportion we see all over the Universe and the prediction of which is touted as one of the great triumphs of the Big Bang model.

2. Hoyle was one of those scientists who was often right when he was wrong. Although his mechanism for making red giants was incorrect, the cold, dense, dark clouds of hydrogen gas he postulated *did* exist. They are the places where new stars are born. Not only that, but 'accretion' – the process by which Hoyle envisioned stars gathering hydrogen gas about themselves – is one of the most important and ubiquitous processes in the Universe. Among other things, it feeds the monster 'supermassive' black holes that lurk at the heart of just about every galaxy, including our own Milky Way.

3. In fact, a star must have a mass of at least three times that of the Sun to ever reach a temperature of 100 million degrees.

CHAPTER 6

1. See Chapter 5.

2. Actually, the build-up of elements inside stars is a little more complicated than this. This is because there are often several alternative routes to building up a particular heavy nucleus. For instance, once carbon-12 and oxygen-16 become common in a star, heavy nuclei can be made by their direct fusion. Thus, two nuclei of carbon-12 can stick to make neon-20 plus a nucleus of helium-4. In practice, the direct fusion of carbon-12 and the direct fusion of oxygen-16 can leapfrog many of the nuclei made by the alpha process.

3. Baade's great discovery was that the stars in the Milky Way fell into two distinct categories. Population I stars, found in the spiral arms of the Galaxy, are dominated by hot, blue stars. Population II stars, in the central 'bulge' of the Galaxy, are dominated by cool, red stars. Later, it would become clear that Population I stars are young and therefore dominated by newborn massive stars. Population II stars are old. All the hot, young stars have gone out, so they are dominated by old, red giant stars.

4. Though a brilliant and visionary astronomer, Zwicky was an eccentric and volatile character whose insistence on calling Baade a Nazi, which he wasn't, eventually led to the shy, quiet Baade living in fear of his life. Zwicky classified people he did not like as bastards or spherical bastards, who were bastards whatever way you looked at them.

5. Every sugar-cube-sized volume would weigh as much as the entire human race. See Chapter 2.

6. The team was at Los Alamos in New Mexico.

7. In fact, the very lightest elements, principally helium, were forged in the Big Bang. The Big Bang model predicts that about 10 per cent of the atoms in the Universe should be helium, forged in the first ten minutes of the Universe's existence. And this is exactly what is observed.

8. Actually, the alpha process makes iron-58 and nickel-62, both of which buck the trend and have slightly less mass per nucleon than iron-56. But the nucleus made by addition of a helium nucleus is zinc-60, and this does have more mass per nucleon.

9. Instead of 26 protons and 30 neutrons, which is the case for iron-56, nickel-56 contains equal numbers of protons and neutrons – 28 of each.

10. An iron meteorite is a natural alloy of stainless steel that contains nickel-56, chromium, magnesium and cobalt, all of which were synthesised in the furnace of a supernova.

11. How are elements heavier than iron made? Well, we know that big, highly charged nuclei like zirconium and uranium cannot be formed by fusing together smaller nuclei because, even if two nuclei hit each other at close to the speed of light – the cosmic speed limit – it would be insufficient to overcome their mutual electrical repulsion. This leaves only processes in which a nucleus captures neutrons, since they have no electrical charge. However, free neutrons disintegrate in about ten minutes. The only way a nucleus can accrue a lot of neutrons is if it is (a) exposed over a short period (less than ten minutes) to an intense burst of neutrons, or (b) exposed over a long period to a source of neutrons which is constantly replenished. The existence of uranium, which is neutron-rich, requires source (a). Such a source is believed to exist in supernovae when the nuclei in the imploding core come apart into neutrons, prior to making a neutron core. The existence of zirconium, which is not neutron-poor, requires (b). With a lot of time available, nuclei would have had time to beta decay, transforming a neutron in their nucleus into a proton. In fact, such a location exists inside normal but highly evolved stars. But iron nuclei will have to soak up a lot of neutrons to turn into nuclei of zirconium or uranium. This is not likely, which explains why zirconium and uranium are rare on Earth.

Further reading:
Home Is Where the Wind Blows by Fred Hoyle (University Science Books, Sausalito, California, 1994).
Supernovae and Nucleosynthesis by David Arnett (Princeton University Press, 1996).

CHAPTER 7

1. Basically, the intensity of starlight from a star drops off with the inverse square of its distance. So if it is twice as far away as a similar star, it is a quarter as bright; if it is three times as far away, it is a ninth as bright; and so on. On the other hand, the volume of a shell of space, which is directly related to the number of stars it contains, increases with the square of its distance. So if it is twice as far away, it contains four times as many stars; three times the distance, nine times as many; and so on. The two factors exactly compensate for each other (at least they do if space is not curved – but that's another story).

2. The speed of light is more than a million times faster than a passenger jet, so you have to admire anyone who finds a way to measure it. Ole Christensen Röemer's idea was to time light crossing a known distance. Since light spanned terrestrial distances too quickly for clocks to measure, the seventeenth-century Danish astronomer looked to the heavens. Imagine there is a clock out in space that strikes midday when the Earth in its orbit around the Sun is closest to the clock. Six months later, when the Earth is at its furthest, the clock will be delayed in striking because the light will have to travel across the diameter of the Earth's orbit. Röemer's genius was to find a celestial 'clock' – Jupiter and its moons. Instead of the striking of midday, he used the instant at which the moon Io went behind Jupiter. In 1676, he found that such 'eclipses' were delayed by 22 minutes (the modern figure is 16 minutes 40 seconds). Combining this with an estimate of the diameter of the Earth's orbit, he calculated the speed of light as 225,000 kilometres per hour. Röemer's measurement was accepted only when confirmed by James Bradley in 1729. His idea was to measure the speed of light relative to something else fast: the speed of the Earth as it orbited the Sun, which he knew. The Earth's motion changed the apparent direction at which light arrived from stars just as your running through rain changes its direction. Bradley measured the shift in position of stars and concluded light travelled at 298,000

kilometres per second, which is almost exactly right.

3. Light actually travels at about 300,000 kilometres per second, or a billion kilometres an hour. Click your fingers. In the time it took you to do that, a ray of light could have made the round trip between Europe and America about 30 times over.

4. The speed of light is only the cosmic speed limit in Einstein's special theory of relativity of 1905. Ten years later, Einstein generalised the theory to deal not only with bodies moving at constant speed with respect to each other but with bodies changing their speed, or accelerating. In his general theory of relativity – which also turned out to be a theory of gravity – space is a backdrop to which the galaxies are effectively nailed. And that backdrop can expand at any speed it likes.

5. The expansion of the Universe also stretches space and, with it, the wavelength of light. Imagine a wiggly line scrawled on a balloon becoming stretched when the balloon is inflated. This 'red shift' – so-called because the stretching shifts visible light to longer, 'redder' wavelengths – results in lower-energy light. It therefore plays a small role in reducing the light energy raining down on the Earth, though it falls far short of explaining why the sky is dark at night.

6. This estimate was made before the 1998 discovery that the Universe's expansion is speeding up. In an ever-growing universe, filling up space with light is like filling up a bath with water as the bath grows bigger at a faster and faster rate.

7. See Chapter 4.

8. The reason for this is that the Big Bang happened *everywhere* in the Universe at once. This is difficult to visualise because, of course, every terrestrial explosion – be it the detonation of a stick of dynamite or a volcano – has a centre. The Big Bang explosion had no centre.

9. The 'peak' emission of the cosmic background radiation is actually at about a millimetre in wavelength, which is not in the microwave 'band'. However, people persist in calling it the cosmic microwave background radiation because the earliest measurements, by Penzias and Wilson and others, were at microwave wavelengths.

Further reading:
Cosmology by Edward Harrison (Cambridge University Press, 1991).

CHAPTER 8
1. It does this because, despite being the weakest force in nature by a

very large factor, it has an infinite range and cannot be screened out. So the more matter there is, the greater its gravity. The gravitational force between a proton and an electron in an atom is 10,000 trillion trillion trillion times weaker than the electric force which keeps matter stiff. So gravity dominates in all bodies which have more than 10,000 trillion trillion trillion atoms, which corresponds to a body about 10 kilometres across. This, incidentally, is why all objects in the Solar System smaller than this are like irregular potatoes, while all objects bigger – like the Earth and the Moon – are crushed by gravity into neat spheres.

2. Quantum theory explains forces as due to the exchange of force-carrying particles. For instance, the electromagnetic force between charged particles arises from the exchange of photons. Think of two tennis players batting a tennis ball back and forth. A force is transmitted to each player by the impact of the tennis ball on their racquet.

3. See Chapter 7.

4. Something that increases exponentially – or is raised to the power of e (about 2.718281828 . . .) – doubles, then doubles in the same time again, and doubles in the same time again, and so on.

5. See Chapter 2.

Further reading:

'The No-Boundary Measure of the Universe' by James Hartle, Stephen Hawking and Thomas Hertog (http://arxiv.org/abs/0711.4630).

CHAPTER 9

1. Incidentally, it follows from the fact that a fly-by and its opposite are indistinguishable that the manoeuvre can never boost the velocity of a space probe relative to the Earth. After all, if it did, you would know which movie was the correct one – the one in which the space probe gained speed. Why, then, does NASA bother? Because while it is perfectly true that a space probe cannot be boosted relative to the Earth, the Earth is moving around the Sun. Consequently, it is possible to choose an ingoing trajectory so that the planet's speed relative to the Sun either adds to or subtracts from the space probe's speed relative to the Earth. Such fly-bys can therefore be used to boost a space probe's speed to reach the planets of the outer Solar System, reduce its speed to reach the planets inside the Earth's orbit, or merely to change its direction. Having said all this, the six spacecraft that have flown past the Earth since 1980 have shown velocity

changes relative to the planet. The origin of these 'fly-by anomalies' is currently a mystery. See 'Anomalous Orbital-Energy Changes Observed During Spacecraft Flybys of Earth' by John Anderson *et al.* (*Physical Review Letters*, Vol. 100, 091102, 2008).

2. Even if you insisted that the laws holding the teacup together are quantum rather than Newtonian, the quantum laws are also time-symmetric. Actually, there is an intrinsic time asymmetry in the law governing nature's weak nuclear force. However, the effect, known as CP violation, is extremely tiny. Also, it seems to have no bearing on processes like the shattering of teacups.

3. Boltzmann's working definition of entropy – synonymous with disorder and randomness – is the 'number of microscopic states possible for a given macroscopic state'. In other words, it is the number of possible ways that the components of an object can be arranged and still yield the object.

4. The light given out by warm bodies, including human bodies, is in the form not of visible photons but invisible 'infrared' photons. Some animals such as pit vipers have organs to 'see' heat, or infrared, enabling them to spot their warm-blooded prey even in the dead of night.

5. Actually, although in the long term the nuclear fuels that replace lost stellar heat become depleted and stars become cold, in the short term they get hotter. In fact, the Sun is about 30 per cent hotter than when it was born. This is because a star is a giant ball of gas. When the gas loses heat, it is no longer able to push outwards as hard against the gravity trying to crush it. The ball shrinks and, in shrinking, is squeezed and heats up. Lewis Carroll knew this. In *Alice in Wonderland*, Tweedledum and Tweedledee pose the riddle: 'What gets hotter as it loses heat?' Answer: a star. (Though I am sure I read this long ago, frustratingly I have not been able to find exactly where and confirm it.)

6. See Chapter 7.

7. Boltzmann, who had also come to the conclusion that the Universe must have been in a special state in the past, speculated that there had been some huge entropy-lowering statistical fluctuation and that our present arrow of time is a consequence of that. Recall that entropy is only overwhelmingly likely to increase. It is not ordained. In highly unlikely circumstances it can decrease. Boltzmann, however, was probably wrong about there being an entropy-lowering event in the past.

8. The nuclei were about 90 per cent hydrogen and about 10 per cent helium, with a tiny sprinkling of other light nuclei such as lithium. All were cooked in the era of 'nucleosynthesis', which took place between about one and ten minutes after the start of the Universe.

9. The Universe also contains invisible, 'dark' matter, which outweighs ordinary matter by a factor of six or seven. It does not interact with photons of light (which is why it is dark) and so is not thought to have been disrupted by their presence. Consequently, it is believed to have started clumping before ordinary matter.

10. A similar argument has recently been proposed by the Oxford mathematician Roger Penrose. It seems that Schulman's idea is 'in the air'.

11. A purist might dispute the connection between cups breaking and what the large-scale Universe is up to. After all, the Earth could have started out in a highly ordered state – the prerequisite for disorder to increase – simply by chance. Boltzmann's explanation of why entropy increases is a statistical thing. It is not ordained that entropy will increase, only overwhelmingly probable. Consequently, an ordered Earth could have arisen as a highly unlikely local 'statistical fluctuation'. But although this might be a plausible explanation for our local arrow of time, it fails to explain why the stars – a more cosmic phenomenon than teacups – are pumping starlight into space and thus continually boosting the entropy of the Universe. They can be doing this only if every one of them started off in a more ordered state, which implies the Universe started off in an ordered state. So here the local is explicitly connected to the cosmic.

Further reading:

Ludwig Boltzmann: The Man Who Trusted Atoms by Carlo Cercignani (Oxford University Press, 1998).

'Sources of the Observed Thermodynamic Arrow' by L. S. Schulman (http://xxx.lanl.gov/abs/0811.2787).

CHAPTER 10

1. The complexity of the everyday world is indeed due to the fact that there are 92 types of atomic building blocks rather than just one, as pointed out in Chapter 3. However, as with many things in science, there is a deeper level of explanation. And that is what we are talking about here – the ultimate source of the complexity of the Universe.

2. Notice the switch from talking about the information needed to *describe* the Universe to the information *contained* in the Universe. The two statements are equivalent. They reflect the growing suspicion among physicists that information is a fundamental 'thing', underpinning all of physics.

3. Binary was invented by the seventeenth-century mathematician Gottfried Leibniz. It is a way of representing numbers as a string of 0s and 1s. Usually, we use decimal, or base 10. The right-hand digit represents the 1s, the next digit the 10s, the next the 10 × 10s, and so on. So, for instance, 9217 means 7 + 1 × 10 + 2 × (10 × 10) + 9 × (10 × 10 × 10). In binary, or base 2, the right-hand digit represents the 1s, the next digit the 2s, the next the 2 × 2s, and so on. So, for instance, 1101 means 1 + 0 × 2 + 1 × (2 × 2) + 1 × (2 × 2 × 2), which in decimal is 13.

4. Usually, the number of distinct arrangements of the subcomponents of a system is defined as $e^{(\text{information content})}$ rather than $2^{(\text{information content})}$, with e, one of the most famous constants in maths, being 2.718281828 . . . The difference is not important since all the figures in this chapter are rough, 'order of magnitude' estimates.

5. Actually, it is a bit more complicated than this and the number of possible arrangements of N thermal photons is e^N. But e^N is approximately the same as 2^N.

6. The reason our 13.7 billion-year-old Universe is 84 billion light years across and not 13.7 × 2 billion light years across, as might naively be expected, is that during inflation, it expanded faster than the speed of light. See Chapter 7.

7. The conservation of information is behind the black hole 'information paradox', highlighted by Stephen Hawking in 1976. A black hole, it turns out, is not completely black but shines with 'Hawking radiation', which allows it to 'evaporate' and eventually disappear. A paradox arises because the Hawking radiation can carry no information about the interior of the black hole, since, by definition, nothing can escape from one. So when the black hole has gone, there remains the puzzle of what happened to the information that described the dying star whose catastrophic shrinkage led to the creation of the black hole in the first place. The strong suspicion now is that it goes into creating myriad tiny bumps on the 'event horizon', the imaginary surface – or point of no return for in-falling matter – that surrounds the black hole. This means

that the information that described the precursor star – a three-dimensional body – is encoded in the two-dimensional event horizon. The event horizon is like a hologram. The implications of this for the Universe are fascinating because it too is surrounded by a horizon – a horizon in time rather than space, but a horizon nonetheless. It appears that the three-dimensional Universe can contain no more information than can be impressed on the two-dimensional surface that surrounds it. Remarkably, it seems we are living in a giant cosmic hologram.

8. $2^{(10^{89})}$ is 2 multiplied by itself ($10 \times 10 \times 10 \ldots$) times, where there are 89 tens inside those brackets. Or, to put it another way, $2^{(10^{89})}$ is approximately $(10^3)^{89} = 10^{267}$.

9. To some extent this was anticipated by the great British physicist Paul Dirac in 1939. Although he knew nothing of inflation and how ridiculously small the Universe had been at the outset, he nevertheless realised that if the Universe was expanding, as the observations of galaxies indicated, it would have been far smaller in the past, which meant it would have been too simple to seed the complexity we see around us today. At least it would be too simple if classical physics described the Universe. Dirac realised, however, that quantum theory might come to the rescue and that unpredictable quantum jumps in the early Universe might be the origin of the Universe's complexity. By recognising the role of quantum theory in the origin of the Universe, Dirac anticipated by four decades the field of quantum cosmology. See *The Strangest Man: The Hidden Life of Paul Dirac, Quantum Genius* by Graham Farmelo (Faber & Faber, 2009).

10. See Chapter 8.

11. Bizarrely, the discovery of the 'dark energy' in 1998 shows that in the past few billion years, the normal vacuum has changed back into an inflationary-type vacuum, though with a tiny, tiny fraction of the energy that drove inflation. Dark energy, like the inflationary vacuum, is speeding up the expansion of the Universe. Nobody knows whether there is any connection between the inflationary phase and the current dark-energy-driven phase, though if there is, two mysteries would be reduced to one.

12. For a long time the vacuum energy was exactly zero, so it contained no energy to dump into other forms even if it were possible for it to decay. However, in the past few billion years, with the arrival on the cosmic stage of the dark energy, everything has changed. Since

nobody knows why the dark energy switched on in the first place, it is always possible that one day it will switch off. However, since the dark energy is so small compared to the vacuum energy that drove inflation, the information injected into the Universe by such a decay will be correspondingly smaller.

Further reading:
'Information, Information Processing and Gravity' by Stephen Hsu (http://arxiv.org/abs/0704.1154).

CHAPTER 11

1. Plutonium-239 is one of two heavy nuclei that split, or 'fission', when struck by a neutron, liberating a large amount of energy. Since further free neutrons are created, more nuclei can be fissioned, causing a runaway nuclear 'chain reaction' and the unleashing of a dam burst of energy. This is an 'atomic bomb'. The other nucleus that can that support a runaway chain reaction is uranium-235.

2. A more trivial 'Fermi problem', typical of the kind Fermi used to challenge his students, was: 'How many piano tuners are there in Chicago?' A rough estimate can be obtained by the following reasoning. Chicago has a population of about 10 million. Pianos tend to be owned by families, not individuals (assuming pianos owned by schools, concert halls and so on account for only a small minority). If an average family has five members, then that makes 2 million families in Chicago. If one in 20 families owns a piano, then there must be 100,000 pianos. Say each piano requires tuning just once a year. That makes 100,000 tunings a year. And say a piano tuner can tune two pianos a day and works about 200 days per year, thus tuning 400 pianos a year. Since there are 100,000 pianos in Chicago, the city must have about 250 piano tuners.

3. The concept of self-reproducing machines would later be explored in detail by the Hungarian-American physicist John von Neumann, famous for inventing the modern computer program. Such probes are therefore often called von Neumann probes.

4. The first person to spell out this argument involving von Neumann self-reproducing space probes was the American physicist Frank Tipler in 1981.

5. Pluto is nowadays not considered a planet but as just one among maybe 100,000 icy 'Kuiper Belt' objects in the outer Solar System. So, essentially, Earth is the only planet with a moon comparable in size to itself.

6. In 1952, Stanley Miller and Harold Urey of the University of Chicago famously took a mixture of gases thought to have existed on the primordial Earth and subjected it to electrical discharges and ultraviolet light. Their experiment yielded aldehydes, carboxylic acids and amino acids, precursors of life. But things stalled there.

7. See my book *The Universe Next Door* (Headline, 2002).

8. See my book *The Never-Ending Days of Being Dead* (Faber & Faber, 2008).

Further reading:

'Explanation of the Code "6EQUJ5" on the Wow! Computer Printout' by Jerry Ehman (http://www.bigear.org/6equj5.htm).

'Scintillation-Induced Intermittency in SETI' by James Cordes, Joseph Lazio and Carl Sagan (*Astrophysical Journal*, Vol. 487, p. 782, 1997).

'When Will We Detect the Extraterrestrials?' by Seth Shostak (*Acta Astronautica*, Vol. 55, p. 753, 2004).

Where Is Everybody? Fifty Solutions to the Fermi Paradox and the Problem of Extraterrestrial Life by Stephen Webb (Praxis Books, New York, 2002).

'Possibility of Life-Sustaining Interstellar Planets' by David Stevenson (*Nature*, Vol. 400, p. 32, 1999).

The Cosmic Connection by Carl Sagan (Cambridge University Press, 2000).

'An Explanation for the Absence of Extraterrestrials on Earth' by Michael Hart (*Quarterly Journal of the Royal Astronomical Society*, Vol. 16, p. 16, 1975).

'The "Great Silence": The Controversy Concerning Extraterrestrial Intelligent Life' by David Brin (*Quarterly Journal of the Royal Astronomical Society*, fall 1983, Vol. 24, p. 283).

'Five or Six Step Scenario for Evolution?' by Brandon Carter (http://arxiv.org/abs/0711.1985).

A New Kind of Science by Stephen Wolfram (http://www.wolfram-science.com/nksonline/toc.html)

Glossary

ABSOLUTE ZERO Lowest temperature attainable. As a body is cooled, its atoms move more and more sluggishly. At absolute zero, equivalent to −273.15 degrees on the Celsius scale, they cease to move altogether. (Actually, this is not entirely true since the Heisenberg uncertainty principle produces a residual jitter even at absolute zero.)

ACCRETION Key astrophysical process in which the gravity of a body sucks in more and more matter from its surroundings. As the matter swirls inwards, like water going down a plug hole, it can create an 'accretion disc'. Friction within the disk heats it up and this is thought to be the source of the prodigious luminosity of powerful galaxies such as 'quasars'. Here the accretion is onto a 'supermassive' black hole, as much as 10 billion times the mass of the Sun.

ALPHA DECAY The spitting out of a high-speed alpha particle by a large, unstable nucleus in an attempt to turn itself into a lighter, stable nucleus.

ALPHA PARTICLE A bound state of two protons and two neutrons – essentially a helium nucleus – which rockets out of an unstable nucleus during radioactive alpha decay.

ALPHA PROCESS The build-up of heavy atomic nuclei inside stars by the addition of alpha particles. It requires a temperature of about a billion degrees.

ANDROMEDA The nearest big galaxy to our own Milky Way, about 2.5 million light years away. Andromeda and the Milky Way are the dominant, big galaxies in a cluster of at least 40 galaxies known as the Local Group.

ANTHROPIC PRINCIPLE The idea that the Universe is the way it is because, if it was not, we would not be here to notice it. In other words, the fact of our existence is an important scientific observation.

ANTIMATTER Term for a large accumulation of antiparticles. Anti-protons, anti-neutrons and positrons can in fact come together to make anti-atoms. And there is nothing in principle to rule out the possibility of anti-stars, anti-planets and anti-life. One of the greatest mysteries of physics is why we appear to live in a universe made solely of matter when the laws of physics seem to predict a pretty much 50/50 mix of matter and antimatter.

ANTIPARTICLE Every subatomic particle has an associated antiparticle with opposite properties, such as electrical charge. For instance, the negatively charged electron is twinned with a positively charged antiparticle known as the positron. When a particle and its antiparticle meet, they self-destruct, or 'annihilate', in a flash of high-energy light, or gamma rays.

ATOM The building block of all normal matter. An atom consists of a nucleus orbited by a cloud of electrons. The positive charge of the nucleus is exactly balanced by the negative charge of the electrons. An atom is about a ten-millionth of a millimetre across.

ATOMIC ENERGY See Nuclear Energy.

ATOMIC NUCLEUS The tight cluster of protons and neutrons (a single proton in the case of hydrogen) at the centre of an atom. The nucleus contains more than 99.9 per cent of the mass of an atom.

BERSERKER Malevolent extraterrestrial machine bent on the destruction of life. Such machines appeared in a number of novels by the science-fiction writer Fred Saberhagen.

BETA DECAY The ejection of a high-speed electron by an unstable atomic nucleus. The nucleus left behind is of an element with one more proton.

BETA RAY An electron ejected during beta decay. The electron does not exist in the nucleus beforehand but is 'created' when a neutron changes into a proton.

BIG BANG The titanic explosion 13.7 billion years ago in which the Universe is thought to have been born. 'Explosion' is actually a misnomer since the Big Bang happened everywhere at once and there was no pre-existing void into which the Universe erupted. Space and time and energy all came into being in the Big Bang.

BIG BANG THEORY The idea that the Universe began in a super-dense, super-hot state 13.7 billion years ago and has been expanding and cooling ever since.

BINARY Way of expressing numbers using only 0s and 1s. In a binary number, the digit at the end represents 1s, the next digit 2s, the next 4s, the next 8s, and so on. So, for instance, $1101 = 1 + (0 \times 2) + (1 \times 4) + (1 \times 8) = 13$.

BLACK HOLE The grossly warped space–time left behind when a massive body's gravity causes it to shrink down to a point. Nothing, not even light, can escape, hence a black hole's blackness. The Universe appears to contain at least two distinct types of black hole: stellar-sized black holes, formed when very massive stars can no longer generate internal heat to counterbalance the gravity trying to crush them, and 'supermassive' black holes. Most galaxies appear to have a supermassive black hole at their heart. They range from millions of times the mass of the Sun in our Milky Way to billions of solar masses in the powerful quasars.

BLACK HOLE INFORMATION PARADOX A difficulty which arises because the laws of physics do not allow the destruction of information, yet, when a black hole disappears, or 'evaporates', the information which described the precursor star appears to vanish for ever. However, some physicists believe that the missing information is encoded in the membrane, or 'event horizon', surrounding the hole. When the black hole evaporates in a hail of Hawking radiation, this radiation then returns the information to the Universe.

BOLTZMANN BRAIN An intelligent observer created spontaneously out in the depths of space by an improbable convulsion of the quantum vacuum.

BOSON A microscopic particle with integer spin – 0 units, 1 unit, 2 units, and so on. By virtue of their spin, such particles are hugely gregarious, participating in collective behaviour that leads to lasers, superfluids and superconductors.

BOYLE'S LAW The observation that the pressure of a gas is inversely proportional to its volume; that is, halving its volume doubles its pressure.

BROWNIAN MOTION The random, jittery motion of a large body under machine-gun bombardment from smaller bodies. The most famous instance is of pollen grains zig-zagging through water as they are repeatedly hit by water molecules. The phenomenon, discovered by botanist Robert Brown in 1827 and triumphantly explained by Einstein in 1905, was powerful proof of the existence of atoms.

CAUSALITY The idea that a cause always precedes an effect. Causality is a much-cherished principle in physics. However, quantum events such as the decay of atoms appear to be effects with no prior cause.

CELLULAR AUTOMATON A simple computer program that takes an input – a pattern of coloured cells – and applies a simple rule to the input to produce an output – another pattern of coloured cells. The key thing is that the output is fed back in as the next input, rather like a snake swallowing its own tail. In the case of a two-colour, adjacent cell, one-dimensional cellular automaton, the input is the pattern of black-and-white cells on one line and the output is the pattern of cells on the next line. Whether a cell in the second line is black or white depends on a rule applied to its two nearest neighbours in the first line. The rule might, for instance, say: 'If a particular cell in the first line has a black square on either side of it, it should turn black in the second line.'

CEPHEID VARIABLE A very luminous yellow star that swells and shrinks periodically. The pulsation period is related to the intrinsic luminosity of the star. This means whenever a Cepheid is observed, its period reveals its true luminosity. A comparison with its apparent luminosity yields its distance. Cepheids have played a key role in measuring the distance to nearby galaxies such as Andromeda.

CHAIN REACTION See Nuclear Chain Reaction.

CHARON Pluto's largest moon.

CHEMICAL BOND The 'glue' that sticks atoms together to make molecules. It involves the sharing, donating or borrowing of the atoms' outer electrons.

CHEMICAL FUEL A fuel such as coal, oil or dynamite. A rearrangement of the electrons in the material's atoms is associated with the liberation of heat energy.

CLASSICAL PHYSICS Non-quantum physics. In effect, all physics before 1900, when the German physicist Max Planck first proposed that energy might come in discrete chunks, or 'quanta'. Einstein was the first to realise that this idea was totally incompatible with all physics that had gone before.

CNO CYCLE The series of nuclear reactions by which stars significantly more massive than the Sun turn hydrogen into helium. It is called a cycle because it comes back to its starting point, in the end recreating the carbon used by the nuclear reactions.

COMET Small icy body – usually mere kilometres across – that orbits a star. Most comets orbit the Sun beyond the outermost planets in an enormous cloud known as the Oort Cloud. Like asteroids, comets are builders' rubble left over from the formation of the planets.

COMPTON EFFECT The recoil of an electron when exposed to high-energy light, just as if the electron is a tiny billiard ball struck by another tiny billiard ball. The effect is a graphic demonstration that light is ultimately made of tiny, bullet-like particles, or photons.

COMPUTATIONAL UNIVERSE The abstract universe of all conceivable computer programs. Since we can be considered computer programs, we – or at least cyber versions of us – exist in the computational universe.

CONDUCTOR A material through which an electrical current can flow.

CONSERVATION LAW Law of physics that expresses the fact that a quantity can never change. For instance, the conservation of energy states that energy can never be created or destroyed, only converted from one form to another. For example, the chemical energy of petrol can be converted into the energy of motion of a car.

CONTRACTION HYPOTHESIS The idea that the Sun stays hot because gravitational energy is constantly being converted into heat energy as it slowly shrinks. The hypothesis is wrong.

COPERNICAN PRINCIPLE The idea that there is nothing special about our position in the Universe, either in space or in time. This is a generalised version of Copernicus's recognition that the Earth is not in a special position at the centre of the Solar System but is just another planet circling the Sun.

COSMIC BACKGROUND RADIATION The 'afterglow' of the Big Bang fireball. Incredibly, it still permeates all of space 13.7 billion years after the event, a tepid microwave radiation corresponding to a temperature of –270°C.

COSMOLOGY The ultimate science, whose subject matter is the origin, evolution and fate of the entire Universe.

COSMOS Another word for Universe.

CRITICAL MASS The threshold mass of a material such as uranium or plutonium necessary to trigger a runaway chain reaction of nuclear fission.

DARK ENERGY Mysterious 'material' with repulsive gravity. Discovered unexpectedly in 1998, it is invisible, fills all of space and appears to be pushing apart the galaxies and so be speeding up the expansion of the Universe. Nobody has much of a clue what it is.

DARK MATTER Matter in the Universe which gives out no light. Astronomers know it exists because the gravity of the invisible stuff bends the paths of visible stars and galaxies as they fly through space. There is at least six times as much dark matter in the Universe as visible matter. The identity of the dark matter is one of the outstanding problems of astronomy.

DECOHERENCE The mechanism that destroys the weird quantum nature of a body, so that, for instance, it appears localised rather than in many different places simultaneously. Decoherence occurs if the outside world gets to 'know' about the body. The knowledge may be taken away by a single photon of light or an air molecule which bounces off the body. Since big bodies like tables are continually struck by photons and air molecules and cannot remain isolated from their surroundings for long, they lose their ability to be in many places at once in a fantastically short period of time – far too short for us to notice.

DEGENERACY PRESSURE The bee-in-a-box-like pressure exerted by electrons squeezed into a small volume of space. A consequence of the Heisenberg uncertainty principle, it arises because a microscopic particle whose location is known very well necessarily has a large uncertainty in its velocity. The degeneracy pressure of electrons prevents white dwarfs shrinking under their own gravity, whereas the degeneracy pressure of neutrons does the same thing for neutron stars.

DENSITY The mass of an object divided by its volume. Air has a low density and iron has a high density.

DETERMINISTIC LAW A law of physics which predicts some aspect of the future with 100 per cent certainty. All classical – that is, non-quantum – laws do this.

DEUTERIUM A rare isotope of hydrogen. Deuterium contains a neutron as well as a proton in its nucleus.

DIMENSION An independent direction in space–time. The familiar world around us has three space dimensions (left–right, forward–backward, up–down) and one of time (past–future). Superstring theory requires the Universe to have six extra space dimensions. These differ radically from the other dimensions because they are rolled up very small.

DOUBLE-SLIT EXPERIMENT Experiment in which particles are shot at a screen with two closely spaced, parallel slits cut in it. On the far side of the screen, the particles mingle, or 'interfere', with each other to produce a characteristic 'interference pattern' on a second screen. The bizarre thing is that the pattern forms even if the particles are shot at the slits one at a time, with long gaps between – in other words, when there is no possibility of them mingling with each other. This result, claimed Richard Feynman, highlighted the 'central mystery' of quantum theory.

ELECTRIC CHARGE A property of microscopic particles which comes in two types – positive and negative. Electrons, for instance, carry a negative charge and protons a positive charge. Particles with the same charge repel each other, while particles with opposite charge attract.

ELECTRIC CURRENT A river of charged particles, usually electrons, which flows through a conductor.

ELECTRIC FIELD The field of force which surrounds an electric charge.

ELECTROMAGNETIC FORCE One of the four fundamental forces of nature. It is responsible for gluing together all ordinary matter, including the atoms in our bodies and those in the rocks beneath our feet.

ELECTROMAGNETIC WAVE A wave that consists of alternating electric and magnetic fields transverse to its direction of motion. An electromagnetic wave is generated by a vibrating electric charge and travels through space at the speed of light.

ELECTRON Negatively charged subatomic particle typically found orbiting the nucleus of an atom. As far as anyone can tell, it is a truly elementary particle, incapable of being subdivided.

ELEMENT A substance which cannot be reduced any further by chemical means. All atoms of a given element possess the same number of protons in their nucleus. For instance, all atoms of hydrogen have one proton, all atoms of chlorine 17, and so on.

ENERGY A quantity which is almost impossible to define. Energy can never be created or destroyed, only converted from one form to another. Among the many familiar forms are heat energy, energy of motion, electrical energy, sound energy, and so on.

ENERGY, CONSERVATION OF Principle that energy can never be created or destroyed, only converted from one form to another.

ENTANGLEMENT The intermingling of two or more microscopic

particles so that they lose their individuality and in many ways behave as a single entity.

ENTROPY The degree of disorder of a system. More specifically, it is the number of possible ways that the components of a system can be arranged and still yield the object.

EPOCH OF LAST SCATTERING The period, about 380,000 years after the beginning of the Universe, when the fireball of the Big Bang had cooled sufficiently for electrons and nuclei to combine to form the first atoms. Since free electrons are very good at redirecting, or 'scattering', photons, before this time light could not travel in a straight line and the Universe was opaque. Once the electrons were mopped up by atoms, it was possible for photons to travel unhindered in straight lines and the Universe became transparent. Today, we pick up photons from this epoch, greatly cooled by the expansion of the Universe over the past 13.7 billion years, as the cosmic background radiation.

EVENT HORIZON The one-way 'membrane' that surrounds a black hole. Anything that falls through, whether matter or light, can never get out again.

EXPANDING UNIVERSE The fleeing of the galaxies from each other in the aftermath of the Big Bang.

FERMI PARADOX The mystery, highlighted by the Italian-American physicist Enrico Fermi, of why, if intelligence has arisen elsewhere in our Galaxy, it hasn't come here to Earth. 'Where is everybody?' asked Fermi.

FERMI PROBLEM A back-of-the-envelope calculation for which Enrico Fermi was well known.

FERMION A microscopic particle with half-integer spin – ½ unit, ³⁄₂ units, ⁵⁄₂ units, and so on. By virtue of their spin, such particles shun each other. Their unsociability is the reason why atoms exist and the ground beneath our feet is solid.

FLY-BY ANOMALY The mystery of why the five space probes that have swung by the Earth since 1990 have gained or lost speed relative to the Earth. According to Newton's law of gravity, there should be no change of speed.

FORCE-CARRYING PARTICLE A subatomic particle whose exchange, like a tennis ball being batted back and forth between tennis players, give rise to force. For instance, the exchange of photons gives rise to the electromagnetic force.

FUNDAMENTAL FORCE One of the four basic forces which are believed to underlie all phenomena. The four forces are the gravitational force, the electromagnetic force, the strong force and the weak force. The strong suspicion among physicists is that these forces are actually merely facets of a single superforce. In fact, experiments have already shown the electromagnetic and weak forces to be different sides of the same coin.

FUNDAMENTAL PARTICLE One of the basic building blocks of all matter. Currently, physicists believe there are six different quarks and six different leptons, making a total of 12 truly fundamental particles. The hope is that the quarks will turn out to be merely different faces of the leptons.

FUSION See Nuclear Fusion.

GALAXY One of the building blocks of the Universe. Galaxies are great islands of stars. Our own island, the Milky Way, is spiral in shape and contains about 200 billion stars.

GAMMA RAY The highest-energy form of light, generally produced when an atomic nucleus rearranges itself.

GAS Collection of atoms that fly about through space like a swarm of tiny bees.

GENERAL THEORY OF RELATIVITY Einstein's theory of gravity which shows gravity to be nothing more than the warpage of space–time. The theory incorporates several ideas that were not incorporated in Newton's theory of gravity. One is that nothing, not even gravity, can travel faster than light. Another is that all forms of energy have mass and so are sources of gravity. Among other things, the theory predicted black holes, the expanding Universe and that gravity could bend the path of light.

GRAVITATIONAL FORCE The weakest of the four fundamental forces of nature. Gravity is approximately described by Newton's universal law of gravity but more accurately by Einstein's theory of gravity – the general theory of relativity. General relativity breaks down over the singularity at the heart of a black hole and the singularity at the birth of the Universe. Physicists are currently looking for a better description of gravity. The theory, already dubbed quantum gravity, will explain gravity in terms of the exchange of particles called gravitons.

GRAVITATIONAL POTENTIAL ENERGY The energy a mass possesses by virtue of its position in a gravitational field. For instance, a loose slate on a roof is said to have more potential energy than one on the ground. If it falls to the ground, the potential energy is converted into other forms – in the first instance, energy of motion.

GRAVITATIONAL WAVE A ripple spreading out through space–time. Gravitational waves are generated by violent motions of mass, such as the merger of black holes. Because they are weak, they have not been detected directly yet.

GRAVITY See Gravitational Force.

HALF-LIFE The time it takes half the nuclei in a radioactive sample to disintegrate. After one half-life, half the atoms will be left; after two half-lives, a quarter; after three, an eighth; and so on. Half-lives can range from the merest split second to many billions of years.

HAWKING RADIATION The heat radiation which is generated near the event horizon of a black hole. A consequence of quantum theory, it arises because pairs of virtual particles and their antiparticles are continually popping in and out of existence in the vacuum, as permitted by the Heisenberg uncertainty principle. Near the horizon of a black hole, however, it is possible for one particle of a pair to fall into the hole. The left-behind particle, with no partner to annihilate with, is boosted from a virtual particle to a real particle. Such particles stream away from a black hole – although admittedly the effect is small for a stellar black hole – as radiation with a characteristic temperature.

HEAT DEATH The hypothetical state in which all temperature differences in the Universe have been ironed out so that all activity dwindles to a standstill.

HEISENBERG UNCERTAINTY PRINCIPLE A principle of quantum theory stating that there are pairs of quantities, such as a particle's location and speed, that cannot simultaneously be known with absolute precision. The uncertainty principle puts a limit on how well the product of such a pair of quantities can be known. In practice, this means that if the momentum of a particle is known precisely, it is impossible to have any idea where the particle is. Conversely, if the location is known with certainty, the particle's momentum is unknown. By limiting what we can know, the Heisenberg uncertainty principle imposes a 'fuzziness' on nature. If we look too closely, everything blurs like a newspaper picture dissolving into dots.

HELIUM Second-lightest element in nature and the only one to have been discovered on the Sun before it was discovered on the Earth.

Helium is the second most common element in the Universe after hydrogen, accounting for about 10 per cent of all atoms.

HORIZON See Light Horizon, Cosmic.

HORIZON PROBLEM The problem arising from the fact that far-flung parts of the Universe which could never have been in contact with each other, even in the Big Bang, have almost identical properties, such as density and temperature. Technically, they were always beyond each other's horizon. The theory of inflation provides a way for such regions to have been in contact in the Big Bang and so can potentially solve the horizon problem.

HYDROGEN The lightest element in nature. A hydrogen atom consists of a single proton orbited by a single electron. Close to 90 per cent of all atoms in the Universe are hydrogen atoms.

HYDROGEN BURNING The fusion of hydrogen into helium, accompanied by the liberation of large quantities of nuclear binding energy. This is the power source of the Sun and most stars.

HYDROSTATIC EQUILIBRIUM The state in which the gravitational force trying to crush a star is perfectly balanced by the force of its hot gas pushing outwards.

INFLATION, THEORY OF The idea that in the first split second of creation the Universe underwent a fantastically fast expansion. In a sense inflation preceded the conventional Big Bang explosion. If the Big Bang is likened to the explosion of a grenade, inflation was like the explosion of an H-bomb. Inflation can solve some problems with the Big Bang theory, such as the horizon problem.

INFLATON The hypothetical force field that drove inflation.

INFRARED Type of invisible light given out by warm bodies.

INSTANTANEOUS INFLUENCE See Non-Locality.

INTERFERENCE The ability of two waves passing through each other to mingle, reinforcing where their peaks coincide and cancelling where the peaks of one coincide with the troughs of another.

INTERFERENCE PATTERN Pattern of light and dark stripes which appears on a screen illuminated by light from two sources. The pattern is due to the light from the two sources reinforcing at some places on the screen and cancelling at others.

INTERSTELLAR MEDIUM The tenuous gas and dust floating between the stars. In the vicinity of the Sun this gas comprises about one hydrogen atom in every three cubic centimetres, making it a vacuum far better than anything achievable on the Earth.

INTERSTELLAR PLANET Hypothetical planet that wanders alone in the deep freeze of interstellar space. Such planets could have been ejected from the vicinity of stars during the process of planet formation.

INTERSTELLAR SCINTILLATION The interstellar equivalent of twinkling. Just as starlight is affected by turbulence in the atmosphere, radio waves from distant celestial objects are affected by turbulence in the interstellar medium. This can cause the radio signal to fluctuate in brightness.

INTERSTELLAR SPACE The space between the stars.

ION An atom or molecule which has been stripped of one or more of its orbiting electrons and so has a net positive electrical charge.

ISOTOPE One possible form of an element. Isotopes are distinguishable by their differing masses. For instance, chlorine comes in two stable isotopes, with masses of 35 and 37. The mass difference is due to the differing number of neutrons in their nuclei. For instance, chlorine-35 contains 18 neutrons, and chlorine-37 contains 20. (Both contain the same number of protons – 17 – since this determines the identity of an element.)

KUIPER BELT A belt of icy bodies orbiting in the outer Solar System. There may be tens of thousands of them. Pluto is now recognised to be one of the larger – though not the largest – Kuiper Belt object.

LASER Light source in which the gregarious nature of photons, which are bosons, comes to the fore. Specifically, the more photons there are passing through a material, the greater the probability that other atoms will emit others with the same properties. The result is an avalanche of photons all travelling lockstep.

LIGHT, CONSTANCY OF The peculiarity that in our Universe the speed of light in empty space is always the same, irrespective of the speed of the source of light or of anyone observing the light. This is one of two cornerstones of Einstein's special theory of relativity, the other being the principle of relativity.

LIGHT, SPEED OF The cosmic speed limit – 300,000 kilometres per second.

LIGHT FILL-UP TIME The time required for stars to fill empty space with light – pretty much like water filling a bath – so that the night sky would appear bright rather than dark. In fact, this time is far greater than the average lifetime of stars.

LIGHT HORIZON, COSMIC The Universe has a horizon much like the horizon that surrounds a ship at sea. The reason for the Universe's horizon is that light has a finite speed and the Universe has been in existence for only a finite time. This means that we only see objects whose light has had time to reach us since the Big Bang. The observable universe is therefore like a bubble centred on the Earth, with the horizon being the surface of the bubble. Every day the Universe gets older (by one day), so every day the horizon expands outwards and new things become visible, just like ships coming over the horizon at sea.

LIGHT YEAR Convenient unit for expressing the distances in the Universe. It is simply the distance light travels in one year, which turns out to be 9.46 trillion kilometres.

LUMINOSITY The total amount of light pumped into space each second by a celestial body such as a star.

MAGNETIC FIELD The field of force which surrounds a magnet.

MANY WORLDS The idea that quantum theory describes everything, not simply the microscopic world of atoms and their constituents. Since quantum theory permits an atom to be in two places at once, this must mean that a table can be in two places at once. According to the Many Worlds scenario, however, the mind of the person observing the table splits into two, one which perceives the table in one place and another which perceives it in another. The two minds exist in separate realities, or universes.

MASS A measure of the amount of matter in a body. Mass is the most concentrated form of energy. A single gram contains the same amount of energy as 100,000 tonnes of dynamite.

MASS SPECTROGRAPH Device for comparing the masses of atoms – or, to be more precise, ions. It does this by measuring how much a magnetic field bends the trajectory of an atom flying through space. The more massive the atom, the less its path is bent.

MAXWELL'S EQUATIONS OF ELECTROMAGNETISM The handful of elegant equations, written down by James Clerk Maxwell in 1868, which neatly summarise all electrical and magnetic phenomena. The equations reveal that light is an electromagnetic wave.

METEORITIC HYPOTHESIS The idea that the Sun is kept hot by a constant raining down of meteorites onto its surface. Unfortunately, it is wrong.

MILKY WAY Our Galaxy.

MOLECULE Collection of atoms glued together by electromagnetic forces. One atom, carbon, can link with itself and other atoms to make a huge number of molecules. For this reason, chemists divide molecules into 'organic' – those based on carbon – and 'inorganic' – the rest.

MOMENTUM The momentum of a body is a measure of how much effort is required to stop it. For instance, an oil tanker, even though it may be going at only a few kilometres an hour, is far harder to stop than a Formula 1 racing car going at 200 kilometres per hour. We say the oil tanker has more momentum.

MOMENTUM, CONSERVATION OF Principle that momentum can never be created or destroyed.

MULTIVERSE Hypothetical enlargement of the cosmos in which our Universe turns out to be one among an enormous number of separate and distinct universes. Most universes are dead and uninteresting. Only in a tiny subset do the laws of physics promote the emergence of stars and planets and life.

NEUTRINO Neutral subatomic particle with a very small mass that travels very close to the speed of light. Neutrinos hardly ever interact with matter. However, when created in huge numbers they can blow a star apart, as in a supernova.

NEUTRON One of the two main building blocks of the atomic nucleus at the centre of atoms. Neutrons have essentially the same mass as protons but carry no electrical charge. They are unstable outside of a nucleus and disintegrate in about ten minutes.

NEUTRON STAR A star that has shrunk under its own gravity to such an extent that most of its material has been compressed into neutrons. Typically, such a star is only 20 to 30 kilometres across. A sugar cube of neutron-star stuff would weigh as much as the entire human race.

NEWTON'S UNIVERSAL LAW OF GRAVITY The idea that all bodies pull on each other across space with a force which depends on the product of their individual masses and the inverse square of their distance apart. In other words, if the distance between the bodies is doubled, the force becomes four times weaker; if it is tripled, nine times weaker; and so on. Newton's theory of gravity is perfectly good for everyday applications but turns out to be an approximation. Einstein provided an improvement in the general theory of relativity.

NON-LOCALITY The spooky ability of objects subject to quantum theory to continue to 'know' about each other's state even when separated by a large distance.

NOVA Close binary star system in which one star is a super-dense white dwarf. Matter sucked from the other star spirals down to the white dwarf and, when enough has accumulated, can trigger an orgy of heat-generating nuclear reactions and an explosion.

NUCLEAR CHAIN REACTION An event triggered when an unstable nucleus such as uranium-235 splits, or 'fissions', releasing energy and neutrons, which split further uranium nuclei, releasing more energy and more neutrons, and so on. If such a chain reaction is controlled, the result is a nuclear reactor. If it is an uncontrolled runaway chain reaction, the result is an atomic bomb.

NUCLEAR ENERGY The excess energy released when one atomic nucleus changes into another atomic nucleus.

NUCLEAR FUSION The welding together of two light nuclei to make a heavier nucleus, a process which results in the liberation of nuclear binding energy. The most important fusion process for human beings is the gluing together of hydrogen nuclei to make helium in the core of the Sun, since its byproduct is sunlight.

NUCLEAR REACTION Any process which converts one type of atomic nucleus into another type of atomic nucleus.

NUCLEAR STATISTICAL EQUILIBRIUM Situation in which nuclear reactions are so fast and furious that a steady state is reached in which nuclear reactions building up each nucleus are perfectly balanced by reactions destroying it. Under such conditions, the mix of element 'freezes out' – that is, it does not change with time.

NUCLEON Umbrella term used for protons and neutrons, the two building blocks of the atomic nucleus.

NUCLEOSYNTHESIS The gradual build-up of heavy elements from light elements, either in the Big Bang – Big Bang nucleosynthesis – or inside stars – stellar nucleosynthesis.

NUCLEUS See Atomic Nucleus.

OORT CLOUD A cloud of comets thought to orbit the Sun beyond the orbit of the outermost planet. Estimates put the total number of comets at up to 100 billion.

PANSPERMIA The idea that the seeds of life spread across space from planetary system to planetary system and that simple life on Earth was therefore 'seeded' from the stars.

PAULI EXCLUSION PRINCIPLE The prohibition on two microscopic particles (fermions) sharing the same quantum state. The Pauli exclusion principle stops electrons, which are fermions, piling on top of each other and, consequently, explains the existence of different atoms and the variety of the world around us.

PHOTOCELL A practical device which exploits the photoelectric effect. The interruption of an electric current when a body breaks the light beam falling on a metal can be used to control something, such as an automatic door at the entrance to a supermarket.

PHOTOELECTRIC EFFECT The ejection of electrons from the surface of a metal by photons striking the metal.

PHOTON Particle of light.

PHYSICS, LAWS OF The fundamental laws which orchestrate the behaviour of the Universe.

PLANCK ENERGY The super-high energy at which gravity becomes comparable in strength to the other fundamental forces of nature.

PLANCK LENGTH The fantastically tiny length scale at which gravity becomes comparable in strength to the other fundamental forces of nature. The Planck length is a trillion trillion times smaller than an atom. It corresponds to the Planck energy. Small distances are synonymous with high energies because of the wave nature of matter.

PLANCK TEMPERATURE The super-hot temperature which corresponds to the Planck energy at which gravity becomes comparable in strength to the other fundamental forces of nature.

PLASMA An electrically charged gas of ions and electrons.

PLUM PUDDING MODEL Early model of the atom in which negatively charged electrons were imagined to be spread throughout a diffuse cloud of positive charge like raisins in a plum pudding.

PLUTO Until recently, the outermost planet. Now it has been demoted to 'dwarf planet' status and is considered to be just one of many tens of thousands of icy bodies in the outer Solar System known as Kuiper Belt objects.

PLUTONIUM Element 94. The nucleus of this man-made element can undergo nuclear fission and sustain a runaway chain reaction, liberating a vast amount of nuclear energy. It can therefore make an atomic bomb.

POPULATION I STAR Hot, blue stars found in the Milky Way's 'spiral arms'. Population I stars are young and relatively rich in heavy elements.

POPULATION II STAR Cool, red stars found in the central region of the Milky Way. Population II stars are old and relatively poor in heavy elements.

POSITRON Antiparticle of the electron.

PRISM Wedge of glass or some other dense, transparent medium. Since light of different colour travels at different speed in such a medium, on passing through a prism the colours of white light fan out to form a rainbow-like spectrum.

PROTON One of the two main building blocks of the nucleus. Protons carry a positive electrical charge, equal and opposite to that of electrons.

PROTON–PROTON CHAIN The chain of nuclear reactions by which stars up to about one and a half times the mass of the Sun turn hydrogen into helium.

PULSAR A rapidly rotating neutron star which sweeps an intense beam of radio waves around the sky, much like a lighthouse.

QED See Quantum Electrodynamics.

QUANTUM The smallest chunk into which something can be divided. Photons, for instance, are quanta of the electromagnetic field.

QUANTUM COSMOLOGY Quantum theory applied to the whole Universe. Since the Universe was once smaller than an atom, such a theory is necessary to try and understand the birth of the Universe in the Big Bang.

QUANTUM ELECTRODYNAMICS The theory of how light interacts with matter. It explains almost everything about the everyday world, from why the ground beneath our feet is solid to how a laser works, from the chemistry of metabolism to the operation of computers.

QUANTUM FLUCTUATION The appearance of energy out of the vacuum as permitted by the Heisenberg uncertainty principle. Usually the energy is in the form of virtual particles.

QUANTUM INDISTINGUISHABILITY The inability to distinguish between two quantum events. They may be indistinguishable because they involve identical particles, for instance, or simply because the events are not observed. The crucial thing, however, is that the probability waves associated with indistinguishable events interfere. This leads to all manner of quantum phenomena.

QUANTUM NUMBER A number which specifies a microscopic property which comes in chunks, such as spin or the orbital energy of an electron.

QUANTUM PROBABILITY The chance, or probability, of a microscopic event. Although nature prohibits us from knowing things with certainty, it nevertheless permits us to know the probabilities with certainty.

QUANTUM SUPERPOSITION A situation in which a quantum object such as an atom is in more than one state at a time. It might, for instance, be in many places simultaneously. It is the interaction, or 'interference', between the individual states in the superposition which is the basis of all quantum weirdness. Decoherence prevents such interaction and therefore destroys quantum behaviour.

QUANTUM THEORY Essentially, the theory of the microscopic world of atoms and their constituents. Those who favour the Many Worlds interpretation believe it also describes the large-scale world.

QUANTUM TUNNELLING The apparently miraculous ability of microscopic particles to escape their prisons. For instance, an alpha particle can tunnel through the barrier penning it in the nucleus, the equivalent of a high-jumper jumping a four-metre-high wall. Tunnelling is yet another consequence of the wave-like character of microscopic particles.

QUANTUM UNPREDICTABILITY The unpredictability of microscopic particles. Their behaviour is unpredictable even in principle. Contrast this with the unpredictability of a coin toss. It is unpredictable only in practice. In principle, if we knew the shape of the coin, the force exerted on it, the air currents around it, and so on, we could predict the outcome.

QUANTUM VACUUM The quantum picture of empty space. Far from empty, it seethes with ultra-short-lived microscopic particles which are permitted by the Heisenberg uncertainty principle to blink into existence and blink out again.

RADIOACTIVE DATING The use of the radioactive disintegration of an element such as uranium to date a rock. In practice, unstable uranium ultimately decays into stable lead, so as time passes the proportion of lead to uranium increases, and this can be used to date the material.

RADIOACTIVE DECAY The disintegration of unstable heavy atoms into lighter, stabler atoms. The process is accompanied by the emission of either alpha particles, beta particles or gamma rays.

RADIOACTIVITY Property of atoms which undergo radioactive decay.

RADIUM Highly unstable, or radioactive, element, discovered by Marie Curie in 1898.

RED DWARF Star less massive than the Sun which glows like a dying ember. About 70 per cent of the stars in the solar neighbourhood are red dwarfs, exploding the myth that the Sun is a typical star. In fact, it is more massive, and therefore more luminous, than the average star.

RED GIANT A star which has exhausted the energy-generating hydrogen fuel in its core. Paradoxically, the shrinkage of the core – which is deficient in heat to hold it up against gravity – heats up the interior of the star. Furious burning of hydrogen in a ring of fire around the core causes the outer envelope of the star to balloon up and cool to a dull red colour. A red giant – the future of the Sun – often pumps out about 10,000 times as much heat as the Sun, principally because of its enormous surface area.

RED SHIFT The loss of energy of light caused by the expansion of the Universe. The effect can be visualised by drawing a wiggly light wave on a balloon and inflating it. The wave becomes stretched out. Since red light has a longer wavelength than blue light, astronomers talk of the cosmological red shift. (A red shift can also be caused by the Doppler effect when a body emitting light is flying away from us. It can also be caused when light loses energy climbing out of the strong gravity of a compact star such a white dwarf, something which is known as a gravitational red shift.)

RELATIVITY, GENERAL THEORY OF Einstein's generalisation of his special theory of relativity. General relativity relates what one person sees when they look at another person accelerating relative to them. Because acceleration and gravity are indistinguishable – the principle of equivalence – general relativity is also a theory of gravity.

RELATIVITY, PRINCIPLE OF The observation that all the laws of physics are the same for observers moving at constant speed with respect to each other.

RELATIVITY, SPECIAL THEORY OF Einstein's theory which relates what one person sees when they look at another person moving at constant speed relative to them. It reveals, among other things, that the moving person appears to shrink in the direction of their motion, while their time slows down, effects which become ever more marked as they approach the speed of light.

SCANNING TUNNELLING MICROSCOPE (STM) A device which drags an ultra-fine needle across the surface of a material and converts the up-and-down motion of the needle into an image of the atomic landscape of the surface.

SCHRÖDINGER EQUATION Equation which governs the way in which the probability wave, or wave function, describing, say, a particle changes with time.

SELF-REPRODUCING SPACE PROBE See Von Neumann Probe.

SETI The Search for Extraterrestrial Intelligence. Most searches involve looking for radio signals from other stars, though optical searches are being undertaken as well.

SILICON BURNING The fast and furious chain of element-building reactions that ensues once a massive star makes silicon. In only a day they turn silicon into iron and nickel, the endpoint of normal stellar nuclear reactions. The star is then on the brink of catastrophe and primed to explode as a supernova.

SOLAR SYSTEM The Sun and its family of planets, moons, comets and other assorted rubble.

SPACE–TIME In the general theory of relativity, space and time are seen to be essentially the same thing. They are therefore treated as a single entity – space–time. It is the warpage of space–time that turns out to be gravity.

SPECTRAL LINE Atoms and molecules absorb and give out light at characteristic wavelengths. If they swallow more light than they emit, the result is a dark line in the spectrum of a celestial object. Conversely, if they emit more than they swallow, the result is a bright line.

SPECTRUM The separation of light into its constituent 'rainbow' colours.

SPIN Quantity with no everyday analogue. Loosely speaking, sub-atomic particles with spin behave as if they are tiny spinning tops (only they are not spinning at all).

STAR A giant ball of gas which replenishes the heat it loses to space by means of nuclear energy generated in its core.

STRING THEORY See Superstring Theory.

STRONG NUCLEAR FORCE The powerful short-range force which holds protons and neutrons together in an atomic nucleus.

SUBATOMIC PARTICLE A particle smaller than an atom, such as an electron or a neutron.

SUN The nearest star.

SUPERFORCE Hypothetical force from which each of the four fundamental forces of nature 'froze out' as the Universe cooled in the aftermath of the Big Bang.

SUPERNOVA A cataclysmic explosion of a star. A supernova may, for a short time, outshine an entire galaxy of 100 billion ordinary stars. It is thought to leave behind a highly compressed neutron star.

SUPERSTRING THEORY Theory which postulates that the fundamental ingredients of the Universe are tiny strings of matter. The strings vibrate in a space–time of ten dimensions. The great pay-off of this idea is that it may be able to unite, or 'unify', quantum theory and the general theory of relativity.

TELEPORTATION The clever use of entanglement to pin down the exact state of a microscopic particle, in apparent violation of what is permitted by the Heisenberg uncertainty principle. This enables the information necessary to reconstruct the state of the particle to be sent to a remote site.

TEMPERATURE The degree of hotness of a body. Related to the energy of motion of the particles that compose it.

THERMODYNAMICS, SECOND LAW OF The decree that entropy can never decrease. This is equivalent to saying that heat can never flow from a cold body to a hot body.

TOTAL ECLIPSE OF THE SUN The coverage of the Sun by the disc of the Moon when the Moon moves between the Sun and the Earth.

TRIPLE-ALPHA PROCESS The unlikely process by which stars weld three helium nuclei into a nucleus of carbon, opening the way to the build-up of all heavy elements.

TUNNELLING, QUANTUM See Quantum Tunnelling.

ULTRAVIOLET Type of invisible light which is given out by very hot bodies. Responsible for sunburn.

UNCERTAINTY PRINCIPLE See Heisenberg Uncertainty Principle

UNIFICATION The idea that at extremely high energy the four fundamental forces of nature are one, united in a single theoretical framework.

UNIVERSE All there is. This is a flexible term which was once used for what we now call the Solar System. Later, it was used for what we call the Milky Way. Now it is used for the sum total of all the galaxies, of which there appear to be about 100 billion within the observable universe.

UNIVERSE, EXPANSION OF The fleeing of the galaxies from each other in the aftermath of the Big Bang.

UNIVERSE, OBSERVABLE All we can see out to the Universe's horizon.

URANIUM The heaviest naturally occurring element.

URANIUM-235 Isotope of uranium that undergoes nuclear fission and which can support a runaway chain reaction, liberating a vast amount of energy. This makes an atomic bomb.

VACUUM, QUANTUM See Quantum Vacuum.

VON NEUMANN PROBE A cross between a starship and an intelligent factory. Such a probe would reach a target planetary system and use the resources there to make two copies of itself. In this way such probes could visit every star in the Galaxy in a relatively short time.

WAVE FUNCTION A mathematical entity that contains all that is knowable about a quantum object such as an atom. The wave function changes in time according to the Schrödinger equation.

WAVE–PARTICLE DUALITY The ability of a subatomic particle to behave as a localised billiard-ball-like particle or a spread-out wave.

WAVELENGTH The distance it takes for a wave to go through a complete oscillation cycle.

WEAK NUCLEAR FORCE The second force experienced by protons and neutrons in an atomic nucleus, the other being the strong nuclear force. The weak nuclear force can convert a neutron into a proton and so is involved in beta decay.

WHITE DWARF A star which has run out of fuel and which gravity has compressed until it is about the size of the Earth. A white dwarf is supported against further shrinkage by electron degeneracy pressure. A sugar cube of white dwarf material weighs about as much as a family car.

WORMHOLE A tunnel through space–time which connects widely spaced regions and provides a short cut. Though permitted to exist by Einstein's theory of gravity, a wormhole is unstable and would snap shut in a split second unless held open by a type of matter with repulsive gravity. Nobody knows whether such 'exotic matter' exists in sufficient quantities to make a traversable wormhole.

WOW! SIGNAL A 37-second-long signal picked up by the Ohio State 'Big Ear' radio telescope on 15 August 1977. When astronomer Jerry Ehman saw the off-the-scale pen-recorder trace, he scrawled across it 'Wow!' It never repeated and, to this day, nobody knows whether the Big Ear picked up a real extraterrestrial broadcast.

X-RAY A high-energy form of light.

YOUNG'S DOUBLE-SLIT EXPERIMENT Experiment in which light of a single colour encounters an opaque screen with two closely spaced vertical slits cut into it. On the far side of the screen ripples of light emerge from each slit and overlap with each other. Where the crests coincide, the waves reinforce, enhancing the brightness, and where the crests of one set of ripples coincide with the troughs of the other, the waves cancel, creating dark regions. This 'interference' pattern – a series of alternating dark and bright bands – can be seen on a white screen interposed in the path of the light. The pattern is a defining characteristic of a wave and was seen for the first time by Thomas Young in 1801.

ZOO HYPOTHESIS The idea that the human race, as an emerging intelligence, is off limits to intelligent space-faring races, thus explaining why the Earth has not been visited by extraterrestrials.

Index